WELL LOGGING FOR THE NONTECHNICAL PERSON

Well Logging
for the Nontechnical Person

By David E. Johnson
and Kathryne E. Pile

Copyright © 1988 by
PennWell Publishing Company
1421 South Sheridan/P.O. Box 1260
Tulsa, Oklahoma 74101

Library of Congress cataloging in publication data

Johnson, David E. (David Earl)
   Well logging for the nontechnical person.
   Includes index.
   1. Oil well logging. I. Pile, Kathryne E. II. Title.
TN871.35.J64 1988   622'.18282   88-5851
ISBN 0-87814-329-7

All rights reserved. No part of this book may be
reproduced, stored in a retrieval system, or
transcribed in any form or by any means, electronic
or mechanical, including photocopying and recording,
without the prior written permission of the publisher.

Printed in the United States of America

1 2 3 4 5 92 91 90 89 88

To Jeana and Kathy
    —I got mine published first! D.J.

To Matt, Tom, Dad, and G.L.
    —The four most inspirational men in my life. K.P.

CONTENTS

Acknowledgments x
A Word to the Reader xi
Prologue xiv

1. **Introduction to Logging** 1

    Why do we run logs? 1
    Who uses logs and why? 3

2. **Reading Logs** 7

    Header 7
    Main log section 9
      Vertical scales 9
      Horizontal scales 11
    Inserts 16
    Repeat section 20
    Calibrations 20
    Comprehension check 20

3. **Formation Parameters** 25

    Types of sediments 25
    Porosity 26
      Intergranular porosity 26
      Other types of porosity 30
    Formation analysis 30
      Shaly formations 32
      Reserves estimate 33
    Invasion 33
    Direct vs. indirect measurements 38
    Resistivity 39

4. **Mud Logging** 45

    Rate of penetration and lag 45
      Measuring ROP 45
      Interpreting ROP from the mud log 47
      Lag 48
    Gas detection 48
    Gas detectors 50
    Analyzing returns 51

viii   Contents

   Measuring and recording the readings 52
   Interpretations 54
  Collecting samples 56
   Sample description 57
  Show evaluation 57
   Identification 57
   Porosity 59
   Permeability 59
   Hydrocarbon ratio analysis 60
   Application 61

5. **Resistivity Measurements** 63

  Induction tools 65
  Focused electric logs 70
  Electric logs 74
  Spontaneous potential 76
  Microresistivity tools 78
   Micrologs 79
   Microlaterologs 81
   Microspherically focused logs 81

6. **Porosity Measurements** 83

  Cores 83
  Subatomic interactions 85
  Density logs 89
   Interpretation—density log 90
  Compensated neutron log 100
   Interpretation—compensated neutron log 100
  Sonic log 101
   Interpretation—sonic log 103
  Multiple porosity logs 105
  Quick-and-dirty cross-plot porosity 111
  Gamma-ray logs 113

7. **Putting It All Together** 115

  Questions to ask before reading the log 115
  Reading the log 116
  BVW_{min} quick-look method 117
  Sample reading 118

Contents   ix

8. **Detailed Interpretation**   127

   Sargeant 1-5 example   129
   Gulf Coast example   141

9. **Computer-Generated Logs**   147

   Wellsite computer logs   148
   Kinds of wellsite-computed logs   149
   Computing center logs   153

10. **Specialty Logs**   161

    Dipmeter   161
    Logs for air-drilled holes   164
       Epithermal neutron log   165
       Temperature log   165
       Noise log   165
    Dielectric constant log   165
    Natural gamma-ray spectroscopy log   166
    Formation testing   168
       Drillstem testing   168
       Wireline formation tests   171
    Through-drillpipe logging   171
       Pulsed-neutron logging   172
    Measurement while drilling   173
    Cased-hole logs   173
       Completion logs   173
       Production logs   176
       Through-casing evaluation   178

Afterword   179
Nomenclature   180
Glossary   187
Suggested Reading   193
Index   195

ACKNOWLEDGMENTS

We gratefully acknowledge the help of the major well-logging companies, and in particular Dick Ghiselin and Elton Head of Schlumberger Well Services; Chuck Martin, Anadrill; and Dale Walker, Tyco Production Company.

We especially thank Schlumberger Well Services for their ongoing support and cooperation throughout this project.

A WORD TO THE READER

If you are interested in well logging, or if you work with well logs but don't know much about them, this book is for you. *Well Logging for the Nontechnical Person* is an elementary yet practical text written for nontechnical users of logs—bankers, landmen, geology and engineering technicians, clerks, and secretaries—who come across logs in their daily routines and would like to make sense of those wiggly lines.

Before we begin, though, let's assess your needs. What is your interest in logging? Why do you want to know about logs? How much do you want to know? What are you going to do with the information? Most important, what can you expect to learn from this book? Will it make you an expert in log interpretation, or will it just make you dangerous?

Obviously, only you can answer the first couple of questions. If you have a passing or casual interest in logs, you've come to the right place. Maybe you've found yourself seated across a conference table from someone waving a log in one hand and a cigar in the other, swearing that his log is proof of the opportunity of a lifetime. Or maybe you've filed and handled logs day after day with only a vague idea about what they reveal. You may not want to become a petroleum geologist, engineer, or log analyst. However, you do want a solid background in logging so you can make more competent business decisions.

What will you learn from this book? Well, after reading it you'll be conversant with the main types of logs in use today—mud logs, open-hole and cased-hole wireline logs, computer-generated logs, and measurement-while-drilling logs. You'll be familiar enough with the more common types of open-hole logs to recognize productive zones and wet zones in simple cases. And you'll know where to go for help with the more difficult cases. Though you won't be an expert, you'll know enough to ask the right questions. You'll also know when you have enough information to make a decision and—more important—when you don't.

This book is not the last word on the subject. It is a rudimentary, introductory explanation of a highly complex and technical subject. Use it as such. If you need to make an important decision based on the information contained in a set of logs, get help. Whenever you have money riding on the correct interpretation of a set of logs, seek expert advice from a consulting log analyst or from a log analyst/salesman working for one of the major logging companies.

A second word of caution concerns the use of the logs. We can seldom measure directly the substance we are looking for (oil or gas). Instead,

A Word to the Reader

> All interpretations are opinions based on inferences from electrical or other measurements and we cannot, and do not, guarantee the accuracy or correctness of any interpretations, and we shall not, except in the case of gross or willful negligence on our part, be liable or responsible for any loss, costs, damages or expenses incurred or sustained by anyone resulting from any interpretation made by any of our officers, agents or employees.

Fig. I–1 Escape clause. This type of disclaimer is included on all log interpretations.

we make inferences and best guesses based on sophisticated measurements of other parameters. From these inferences we formulate our interpretations. But think about it for a minute. If logs always succeeded in their predictions, if people interpreted them perfectly, and if logging tools never malfunctioned, logging companies wouldn't need the escape clause that is attached to all of their interpretations (see Fig. I–1).

What is the escape clause? In essence it states that we live in an imperfect world, full of well-intentioned but sometimes inept people; that machines and electrical instruments sometimes fail; and that occasionally interpretations will be wrong. Companies tell you this to underscore the high risk in any drilling venture. Logging companies and log analysts are entirely blameless for any losses incurred as a result of these failures. We echo these sentiments.

A book such as this one, which tries to translate a very technical subject into layman's terms, can never be as precise or exact as a technical treatment of the same subject matter. This book is a compromise between rigorous exactness and oversimplification. We hope we have hit a middle ground where the explanations are correct but simplified. We have often omitted or greatly abbreviated tool design principles and acquisition procedures and have presented only a couple of the many interpretation methods available today. As we stated previously, we are not trying to make log analysts out of our readers; rather, we want to give them an appreciation for the process. A suggested reading list is included in the Appendix for those readers who wish to delve more deeply into this subject.

The book is aimed primarily at the petroleum industry. Oil and gas well logging accounts for most of the logs that are run. Nonetheless, there are several other branches to the logging family tree. A growing segment consists of logging to evaluate mineral deposits for mineral exploration. Logs also play a role in geotechnical engineering, such as the study of the famous San Andreas fault; in environmental impact evaluation and monitoring of waste disposal wells; and in scientific investigation (many logs have been run for the federal government in monitoring and evaluating shot holes for underground atomic weapons testing). The logs used in this type of work are generally the same as those used in petroleum logging; so although your particular application may not be mentioned, this book describes the logs that you might use.

The examples in this book are from the United States—not because they are unique, but because they were handy. An example is just that—an example, illustrating a point. It is not necessary to cover every geologic province in order to apply the methods described here to your particular part of the world. The units of measurement used in the book are those that were used for the log examples: English units. Many logs are run in metric units, but this should be of little importance to the reader because units of measurement are always noted on the log heading and scales.

Regardless of your interest in logs, the part of the world where you want to use them, or the measurement system that you prefer, this book should get you started in understanding petroleum well logs.

David E. Johnson
Kathryne E. Pile

PROLOGUE

One day your neighbor, an independent geologist, comes over to your house and mentions he has a prospect that he's selling to some of his friends. Since you're a neighbor and a good friend, he'll let you in on this hot deal. You're curious, so you invite him to bring over his data.

A few evenings later he arrives at your house, laden with maps, offset logs, and an economic evaluation. On the structure map he points out the proposed location and argues that the well should have 20 ft of pay. Then your friend pulls out the dual-induction and neutron-density logs from the offset well to show you what the porosity and water saturation look like.

As you examine the logs, your friend points out that the zone at 8,720–8,737 ft has at least 15% of cross-plot porosity and an R_t of 45 ohms. He also tells you the formation-water resistivity is 0.04 and the water saturation calculates 20%. The neutron-density curves indicate the formation is gas-bearing. Then your friend looks at you and asks, "What do you think? Want to put some money into it?"

What would you say if this happened to you? Would you be able to make an intelligent evaluation of the formation's potential? If you were able to follow this scenario, you are well acquainted with logs and could make an educated guess. But if you would find yourself floundering over terminology or becoming confused by the squiggles on the logs, you've come to the right place for help. Read on.

1
INTRODUCTION TO LOGGING

Just what are well logs and how did they get their name? One story goes something like this.

When the oil industry was getting started around the close of the last century, many sailors were out of work. (Curiously, the sailors were unemployed because the fledgling oil industry and kerosene were eliminating the need for whale oil.) Since they were used to working at heights and with rigging, they were naturals to scale the tall oilfield derricks.

Along with the influx of sailors came many of their nautical expressions. That's why the drilling derrick and its equipment are called a "rig," the derrick is a "mast," the changing room is called a "doghouse," and the records are kept in the "knowledge box." The term "log" is another of these merchant marine expressions.

Nearly everyone is familiar with the ship's log kept by the captain. It's simply a chronological record of what happens aboard a ship. The record of what occurs on a drilling rig is the driller's log. Oil companies are interested in what happens as a bit drills deeper, so the driller's log is usually recorded by depth rather than by time.

In the early days of the industry, the driller's log was about the only information available on subsurface formations. On it were recorded the types of rock brought up from the borehole, how many feet per hour the bit was drilling, oil or gas flows, equipment breakdowns, accidents such as stuck drillpipe, and any other occurrence that might have a bearing on evaluating the well. Today, "log" has stretched to mean any data recorded vs. depth (or time), in graph form or with accompanying written notes.

When someone mentions a log, he is usually referring to records run on an uncased wellbore using an electric wireline logging truck and tools (Fig. 1–1). Logs can also refer to the driller's log, mud logs, computer-generated logs, and MWD (measurement while drilling) logs.

WHY DO WE RUN LOGS?

What are we trying to accomplish with a log? What does it tell us that is so important?

2 Well Logging for the Nontechnical Person

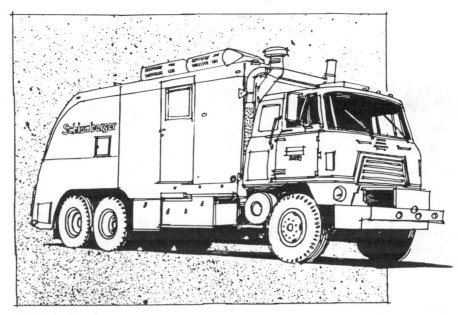

Fig. 1–1 Electric logging truck and array of tools. This modern truck carries an onboard computer and enough cable to log wells as deep as 25,000 ft. (courtesy Schlumberger)

One of the advertising slogans of Schlumberger (pronounced slumber-jay) during the '60s was "... the eyes of the oil industry." This slogan aptly describes the importance of logging. Geologists and engineers literally work blindly when they try to imagine what is happening at the bottom of a well. Layers upon layers of sediments have amassed over the years and have been deformed and altered so much that we can't guess exactly what lies beneath our feet.

Before logging, drillers had only the information from their driller's logs and the behavior of nearby, or offset, wells. This information was and is important and useful, but it still left a lot to "by guess and by golly." Electric wireline logs have turned on the light for the petroleum geologist and engineer. In particular, they provide information in areas such as these:

- depths of formation tops
- thickness of formations
- porosity
- temperature
- types of formations encountered (shale, sandstone, limestone, dolomite)

Introduction to Logging   3

- presence of oil or gas
- estimate of permeability
- reservoir pressures
- formation dip (the angle the formation makes to the horizontal and its direction)
- mineral identification
- bonding of cement to the casing
- amount and kind of flow from different intervals in a producing well

The list goes on and on, and new logs as well as new uses for old logs are being developed continually.

But very simply, the real reason for running logs is to determine whether a well is good or bad. A good well is commercially productive— it produces enough oil or gas to pay back its investors for the cost of drilling and leaves a profit. A bad well is not commercially productive. Logs help us make this determination.

By the time a log can be run, thousands of dollars have been spent for leases, possibly for seismic studies, and for drilling. However, thousands of dollars more are still to be spent to complete the well—running the casing, cementing, perforating, testing, setting production tubing and packers, and installing wellhead equipment and surface production facilities. If a company can determine that a well won't be productive before it spends thousands of dollars on the completion costs, it will minimize its loss. As in poker, there's no sense in throwing good money after bad.

Logs help us determine whether the formation we are penetrating contains commercial reserves of oil or gas, thus minimizing costs on bad wells. On good wells, the logs also show us where the oil or gas may lie, how much there is (reserves), and whether more than one zone is productive.

WHO USES LOGS AND WHY?

Practically everyone in the oil industry uses logs at one time or another (Fig. 1–2). And logs are certainly used by everyone involved in the decision-making processes necessary in drilling and completing a well.

Logs are used in nearly every phase of the exploration and production process. Let's take a look at a deal put together by an independent geologist. First, the exploration or development geologist evaluates an area. He bases his evaluation on seismic data, existing logs, nearby

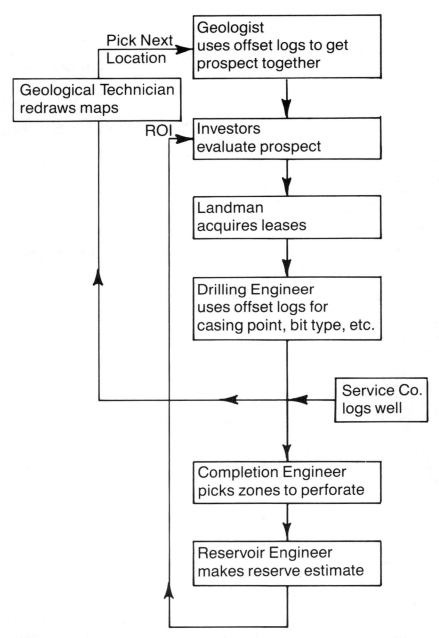

Fig. 1–2 Who uses logs? Logs are used by practically everyone in the industry, as this figure shows.

well data, imagination, and intuition. Armed with this information, the geologist draws structural maps of the area and recommends how to develop the prospect, or "play."

The log might be used next by the banker loaning money or by the investor. These parties have a working interest, i.e., they invest money in the hope that they will reap profits. To protect the investment, bankers or investors often evaluate the log data independently, using in-house or consulting geologists and engineers. When the opinions differ, it's good to know enough about log interpretation to decide which way to go.

The landman is primarily responsible for obtaining the leases necessary for the geologist's play. He doesn't need to know how to read the logs as accurately as a log analyst. However, he must have a working knowledge of logs so he can discuss prospects with landowners, bankers, and geologists.

The drilling engineer drills the well on the basis of log information from nearby wells. From this information, the engineer decides the kind and weight of mud to use, the types of formations to be encountered, the kind of drill bits to use, where to set casing, and how long it will take to drill the well.

The completion engineer relies heavily on logs to determine which zones are probably productive and exactly where the casing should be perforated. On the basis of information from daily reports, the mud log, and various open-hole and cased-hole logs, the completion engineer will perforate, test, treat, and finally put the well on production.

The reservoir engineer uses the open-hole logs to make the initial calculation of reserves (the amount of producible oil or gas). These reserve calculations are updated periodically from production data, pressure buildup tests, and possibly other logs run later in the life of the well.

Included in this string of people who use logs are the geological technician who drafts the structural maps, the royalty owner who wants to know why his well wasn't as good as his neighbor's, the mud salesman making a pitch for a better mud system on the next well, and the accountant who calculates the net worth of his company's assets indirectly using well log data to make his evaluation. Many people depend on the interpretation of well logs. That's why it's good to be as knowledgeable as possible about logs.

The first step toward learning about logs is knowing their components and how to read them, so let's turn to Chapter 2 and get started.

2

READING LOGS

In the next chapters we'll be looking at several different types of logs used in the petroleum industry. Some measure the resistivity of formations, others determine porosity, and still others determine types of minerals present. But before we study what these various logs tell us about the earth's formations and the presence of oil or gas, we need to know where to locate and how to read the five major sections of a log: the header, the main log section, the inserts, the repeat section, and the calibrations.

HEADER

When you are handed a log, the first thing you usually see is a short section of text at the top of the log. This section is called the heading or the header because, as the name implies, it is attached to the top, or head, of the log. The header contains useful and often critical information. As you read the list below, note the corresponding location on Fig. 2–1:

1. Logging company
2. Operating company (operator)
3. Specific well information
   —well name or number
   —lease or field name
   —legal location (where the well is located geographically, often a certain section, township, and range)
   —elevation of ground surface above sea level and usually elevation of the rig floor or the kelly bushing
   —date when the logs were run
   —total depth of the well at the time of logging
   —miscellaneous information such as drilling mud properties, bit size, casing size, and depth
4. Type or kind of log run
5. Other logs or surveys run on the well
6. Equipment information
   —tool serial numbers
   —tool spacings

8   Well Logging for the Nontechnical Person

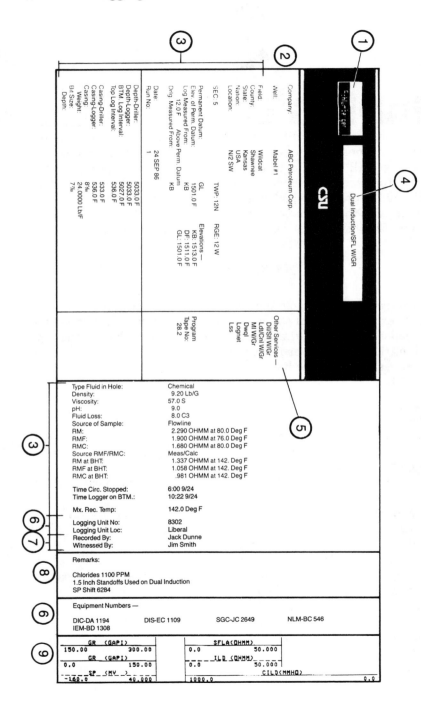

Fig. 2–1 Header. The header provides information on type of well and parameters: (1) logging company, (2) operating company, (3) well, (4) type or kind of log run, (5) other surveys run, (6) equipment, (7) personnel, (8) remarks, and (9) scales and curve identification.

—truck number
—name of office that supplied the truck
7. Personnel information
   —person who recorded the log
   —person (company man) who witnessed the log
8. Remarks section for noting any unusual conditions or occurrences during the logging job
9. Log scales and curve identification

A first rule to follow when interpreting any log is to examine the heading carefully. Why? First of all you can see which logging company ran the log. Not all logging companies are equal in the quality of their work, and some of your decisions may be influenced by how much confidence you have in the log readings. You will also want to look at the well name just to be sure that you're reading the logs from the right well. Another reason to check the heading is to look at mud data (resistivity, water loss, weight, and viscosity), bit size, depth of any casing, and total well depth. All of this information is important to your interpretation of the logs and can help you decide what to do with the well.

MAIN LOG SECTION

Just below the header is the main body of the log, which looks like a very long graph. Here we read the data that the logging equipment transmits to the surface. In this section we must be able to read both vertical and horizontal scales.

Vertical Scales

The vertical or long axis measures the depth of the well and records the exact depth at which formations occur (Fig. 2–2). The depth track or depth column is the vertical space with numbers near the center of the log. Depth numbers are printed in this space in multiples of 100 ft of depth and correspond to horizontal depth lines on the graph.

Depth scales are always linear, that is, the division marks are of uniform size just like the division marks on a ruler. The depth scale on a log is usually either 1, 2, or 5 in./100 ft of hole. This means that if you laid a ruler on the depth scale of Fig. 2–3 and measured between 1,600 and 1,700 ft, the distance would be 2 in. since this is a 2 in./100 ft scale.

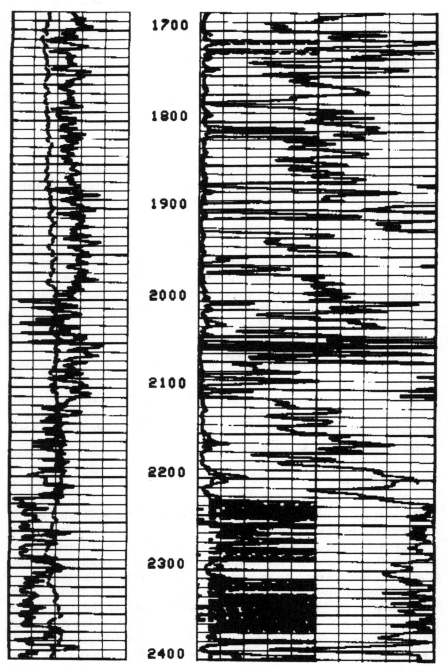

Fig. 2–2 Correlation scale. This vertical scale, which measures 1 in./100 ft, is used to compare formation depths of nearby wells.

In addition to the dark horizontal lines at the 100-ft depth numbers, the 1-in. and 2-in. scales have both 50-ft and 10-ft depth lines. Look again at Fig. 2–3 and find the depth 1,650 ft (A). Now find the depths 1,680 ft (B) and 1,738 ft (C). Note that you had to mentally divide, or interpolate, the distance between the 1,730 and the 1,740 depth lines and estimate where 1,738 is.

The depth lines on the 1 in./100 ft scale are divided the same as on the 2-in. scale: 10-ft, 50-ft, and 100-ft lines and 100-ft depth numbers. If we had a 5-in. log, we would have 100-ft lines at the depth numbers, 50-ft lines of the same weight (degree of darkness or width) as the 100-ft lines, 10-ft lines somewhat thinner than the 50- and 100-ft lines, and 2-ft lines the thinnest of all. By having 2-ft lines and the expanded scale of the 5-in. log, we could easily read depths to 6 in.

The 2 in./100 ft and 1 in./100 ft scales are called correlation scales. Geologists use the correlation scale to compare between several wells over large intervals of formation. The 2-in. scale is usually used to correlate with one or two nearby offset wells, while the 1-in. scale is often used to construct cross sections over several miles of surface and many thousands of feet of formation. The 5 in./100 ft scale is called the detail scale because more features can be noted over 5 inches than over 1 inch.

In addition to these three common depth scales, other special scales are occasionally seen. Super detail scales, 10 in./100 ft or 25 in./100 ft, are used most often with micrologs or fracture identification logs and are run over short intervals of the hole.

To the left of the depth track is track 1 (see Fig. 2–3). This track is often called the SP (spontaneous potential) track or the gamma-ray track after the two curves, or measurements, that are most commonly recorded there. To the right of the depth track are two more measurement tracks, track 2 and track 3. Various kinds of curves are recorded in these two tracks.

Horizontal Scales

We have already talked about the vertical or depth scale. Now let's consider the horizontal scale, which is the measurement scale. It records the changing formation parameters that we are measuring, such as resistivity and porosity. The horizontal scale may take one of several forms, so let's talk about scales and graphs in general and then come back to horizontal log scales and try our hand at reading some.

Fig. 2–4 is a very simple log with measurement curves in tracks 1 and 2. Note that the curves are labeled at the bottom. The curve in

12   Well Logging for the Nontechnical Person

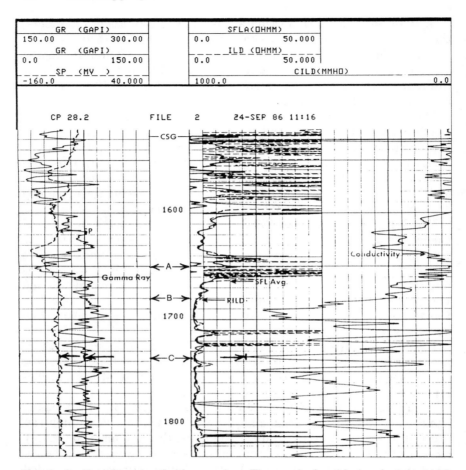

Fig. 2–3 Reading depths from a log. The scale for this log is 2 in./100 ft. For practice in reading logs, identify the depths at points A, B, and C.

track 1 is called GR (gamma ray) and is represented by a solid line that is scaled from 0 on the left, at division 0, to 100 on the right, at division 10. We can determine how many GR units are represented by each division by dividing 100 (the number of units) by 10 (the number of divisions). Each division is worth 10 GR units.

Now look at the curve in track 2. This curve is also solid (it could be short dashed, long dashed, or dotted, and the curve weight could be light or heavy). The curve in track 2 is called R (resistivity) and is scaled from 0 at 0 divisions of track 2 to 20 at 10 divisions of track 2. The R curve has a backup curve, shown by a heavy solid line, that is

Reading Logs 13

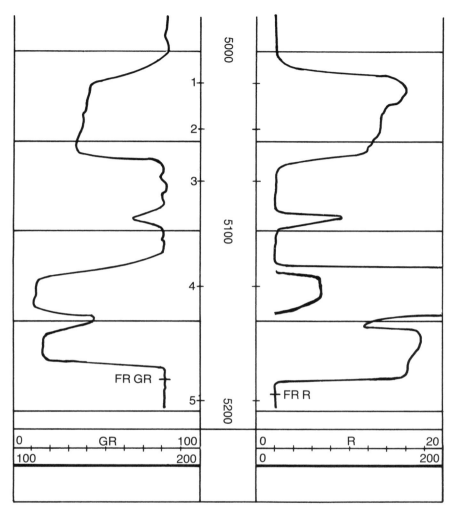

Fig. 2–4 A simple log. Note the two tracks, one scaled into 100 units and the other into 20 units.

scaled from 0 to 200. This curve, a one-tenth backup curve, does not print or show on the log until the primary curve goes off the scale. Note that 10 divisions of track 2 are the same as 0 divisions of track 3. Often curves will be scaled completely across both tracks 2 and 3; in other situations, such as this one, the curve will cut off (disappear) at division 10.

How many R units are there per division? If you said 2 for the primary curve and 20 for the backup curve, you're correct. Now let's practice reading these curves. First, make a table like this on a piece of paper:

Point	Depth	GR	R
1			
2			
3			
4			
5			

Read the depth of each point and the curve values at that point. For the first point we get 5,020 ft. The GR is four divisions from 0, so it must have a value of 4 times 10, or 40. Write 40 under GR next to the depth 5,020. For the R reading, you should have 8 divisions times 2 R units per division, or 16. Go ahead and fill in your table with the other readings. The answers are at the end of this chapter.

Be sure to notice one other point on Fig. 2–4. At the very bottom of the curves is a short line labeled FR GR and FR R. FR stands for the first reading of that particular curve. **Never** read a curve below this point. Even though the curve might appear to be recording a measurement, it's not; the logging tool is resting on the bottom of the well. The reason why the FR R is deeper than the FR GR is that the GR tool was above the R tool on the wireline and could not be lowered as far.

Now look at Fig. 2–5 for some more practice in reading log values. Once again, start by looking at the curve identification at the bottom of the log. Note that track 1 has a GR curve scaled from 0 to 150 with a backup scale of 150 to 300. This means each division will be worth 15 units on the GR scale.

Tracks 2 and 3 are much more complicated than in the last example. Here we have three different curves with different scales and curve codings. The solid curve is labeled RHOB and is scaled from 2.0 at 0 divisions of track 2 to 3.0 at 10 divisions of track 3. (Ten divisions of track 3 are the same as 20 divisions of track 2.) To find out how many units of RHOB each division is worth, subtract the reading at 0 divisions from the reading or scale value at 20 divisions and divide by 20. The answer is 0.05 units per division.

Where is the 0 line for the PHIN curve? To determine this, first you have to find out how many PHIN units there are per division. Note that the scale reads backwards, that is, the higher values are on the left and the lower values are on the right. In fact, division 10 of track

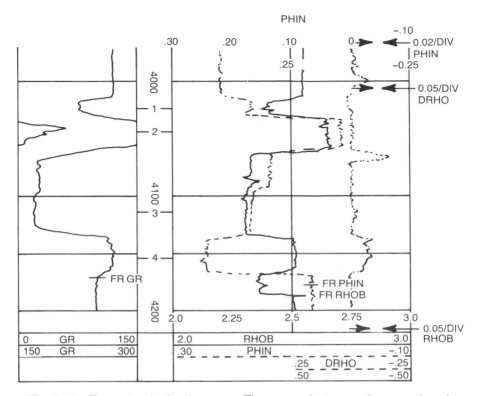

Fig. 2–5 Three-track simple curve. The space between the arrowheads indicates the value of one division for each curve. Try your skills by reading the log at points 1-4.

3 is labeled −0.10, a negative value. How do we determine units per division? We subtract the value at division 0 (0.30) from the value at the right edge of the track, in this case division 20 of track 2 (−0.10). The answer is −0.40 divided by 20, or −0.02 per unit. The minus sign simply means that the scale increases from right to left instead of from left to right. So where is 0 PHIN? It's at division 5 of track 3 (division 15, track 2). The RHOB and DRHO scales are both 0.05 per division.

Now let's get some more log-reading practice by recording the values of GR, PHIN, RHOB, and DRHO at points 1, 2, 3, and 4. Set up a table as you did for the previous exercise.

The horizontal scale of track 1 is always linear, that is, the increments are of uniform size like a ruler's. However, tracks 2 and 3 are often printed in a variety of formats. Besides being linear, both tracks may be logarithmic or the two tracks may even be a split scale in which track 2 is logarithmic and track 3 is linear.

Logarithms are based on powers of 10. That is, each cycle of numbers is 10 times as large as the preceding cycle. Think of logarithms in terms of money (see Fig. 2–6). If the lowest value shown is 1 penny, the first cycle is scaled from 1 cent to 1 dime, or 10 cents. The second cycle is scaled from 1 dime (10 cents) to 1 dollar or 10 dimes or 100 cents. The third cycle is scaled from 1 dollar (100 cents) to 10 dollars (1,000 cents).

On Fig. 2–6 read the values at points 1, 2, 3, and 4. The value at point 1 is 20 cents or $0.20. (See the scales at the bottom of the figure.) Now read the points for 2, 3, and 4. Notice that point 5 is a little different; it shows how a log scale may be used to give a more precise reading of a value. If we try to read just the point at the far right, we estimate something between 500 and 550; but by using the other two points, we can refine our reading to get 525 (500 + 20 + 5).

To get some practice with a real log, note Fig. 2–7. Here we have a dual induction-SFL curve recorded on log scale in tracks 2 and 3. Note that the scale starts at 0.20 instead of 0.10, but otherwise it is just the same as the example in Fig. 2–6. Read the three curves at points 1 and 2; check your answers at the end of the chapter.

Other common presentations record the depth track at the left of the log and divide the space to the right into four tracks. This presentation is often used with computer logs, which are generated from combinations of calculations and/or interpretations from two or more logs.

INSERTS

Between the major sections of the log and the bottom are inserts, which label the measurement scales and identify curves. Curves may be printed as solid, long-dashed, short-dashed, or dotted lines, and all of these lines may be either heavy or light.

For an example, let's look at Fig. 2–8. In the upper part of the insert that labels the curves at the bottom of the 2-in. log scale, the gamma-ray (GR) curve is labeled in track 1 with a scale of 0 to 150 units and is a heavy solid line. The SP curve is labeled as a dashed line with a scale of -160 to 40 (20 units per division). In track 2, the SFLA curve is shown as a solid line with a scale of 0 to 50 ohms. The ILD curve is shown as a dashed line with a scale of 0 to 50 ohms. In tracks 2 and 3 is the CILD curve with a scale of 2,000 to 0 millimhos, shown as a solid line.

Reading Logs 17

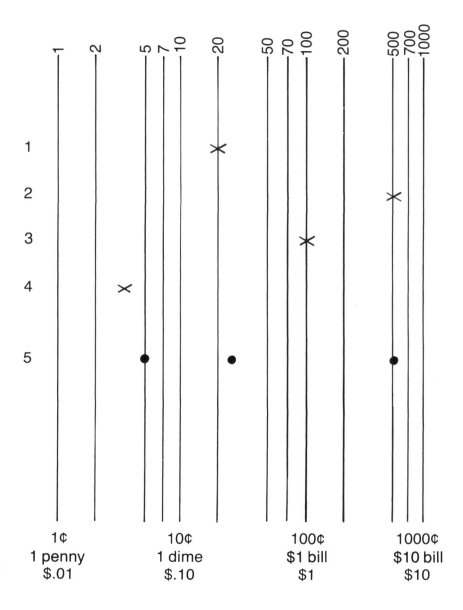

Fig. 2–6 Logarithmic practice. Read the graph using the guidelines in the text to learn how to read logarithmic scales.

18   Well Logging for the Nontechnical Person

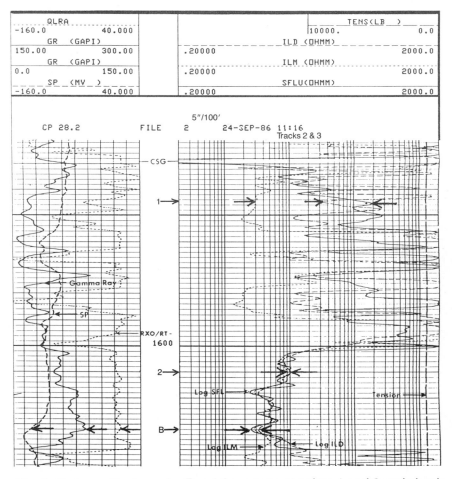

Fig. 2–7 Logarithmic scale. Read the curves at points 1 and 2 and check your answers at the end of the chapter.

The purpose of the insert is to identify all of the curves on each section of the log so the various curves can still be recognized as the presentation changes from the 2 in. to the 5 in./100 ft scale. Note that at the top of the 5-in. log all of the curves are identified again. Here we have some additional curves, and tracks 2 and 3 are now in the logarithmic format. See if you can properly label and locate all of the curves shown for the 5-in. scale.

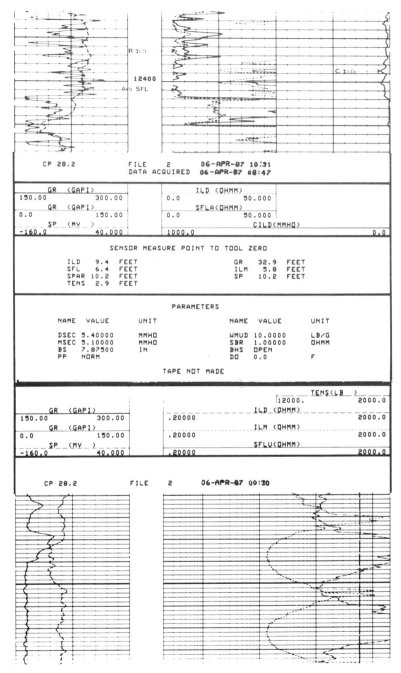

Fig. 2–8 Insert. Inserts list scales and identify curves.

REPEAT SECTION

Below the main body of the log comes the repeat section, which verifies that the logging equipment was working properly and that the measurements repeat themselves (Fig. 2–9). Always lay the repeat section alongside the main log section and check at several different places to make certain the readings repeat. If they do not, rerun the logs with different equipment if possible. If this is not possible, use the log information with caution. Radioactive measurements always show a certain amount of variation because of the random nature of radioactive decay; however, other measurements, such as resistivity and sonic, should repeat very closely.

CALIBRATIONS

The final section of the log shows the before-and-after survey (logging) calibrations. These calibrations verify that the tools were properly adjusted before the log was run and that they were still in adjustment at the end of the logging job. For our purposes, we will assume the tools were all properly calibrated. If you ever suspect a log was not calibrated properly or the tools changed during logging and became uncalibrated (out of tolerance), then contact the logging company, question their procedures, and have them verify that the tools were properly calibrated before and after the survey.

COMPREHENSION CHECK

As a test to see how well you understand this chapter and especially how well you read log values, turn back to Figs. 2–3, 7, and 9 and read all of the curves and depths at points A, B, and C. Check your answers at the end of the chapter. If you don't do too well, go back and reread the sections you had trouble with.

We now know how to identify and read the major components of a log. Now let's see what these curves are telling us so we can put everything together and interpret our logs.

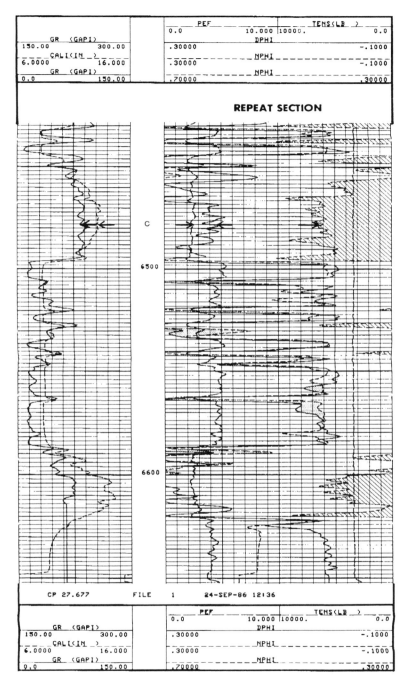

Fig. 2–9 Repeat section. This portion of the log verifies that the logging equipment was working properly and that the measurements repeat themselves.

ANSWERS TO FIGURES 2–4, 2–5, 2–6, 2–7.

Fig. 2–4

Point	Depth	GR	R
1	5020	40	16
2	5043	36	13
3	5075	82	2
4	5131	11	70
5	5194	*	*

*These points are below the FR (first reading). **Never** read a curve here; it's not a valid measurement.

Fig. 2–5

Point	Depth	GR	PHIN	RHOB	DRHO
1	4024	75	0.17	2.40	0.0
2	4045	180	0.015	2.65	0.01
3	4114	20	0.17	2.31	0.02
4	4155	120	0.25	2.52	0.06

Fig. 2–6

Point	Value	Monetary Value
1	20	20¢
2	500	500¢ or $5
3	100	100¢ or $1
4	3	3¢
5	525	525¢ or $5.25

Fig. 2–7

Point	Depth	ILD	ILM	SFLU
1	1545	32	3	200
2	1610	10	9	10

ANSWERS TO COMPREHENSION CHECK

Fig. 2–3

Point	Depth	GR	SP	SFLA	ILD
A	1650	30	−120	150	—
B	1670	75	−92	3	5
C	1737	75	−100	2	2

Fig. 2–7

Point	Depth	GR	SP	ILD	ILM	SFLU
B	1632	85	−130	3.2	2.9	2.5

Fig. 2–9

Point	Depth	GR	CAL	PEF	DPHI	NPHI
C	6475	85	13.6	2.4	20.5	4

3
FORMATION PARAMETERS

Before we can begin interpreting logs, we need to be sure we understand a few basics of reservoir behavior, particularly types of sediments, porosity, formation analysis, indirect measurements, and resistivity.

TYPES OF SEDIMENTS

As you are probably aware, oil and gas deposits are contained within sedimentary rocks, one of the three major types of rock. These rocks, formed by sediments deposited in layers on the bottom of rivers, lakes, and oceans, can be classified into three categories: clastics, evaporites, and organics.

Clastics, from the Greek word *klastos,* meaning "broken," are formed from the fragments of other rocks. The most common example of this type of rock is sandstone. Individual sandstone grains are fragments of other rocks that were weathered, worn, crushed, and tumbled by wind and water until they were deposited, buried by other sediments, and cemented by chemical action, heat, and pressure. Sandstones (or sands, as they are commonly called) are the most important reservoir rock; most hydrocarbon accumulations occur in sandstones.

Clastic sediments are differentiated by the size of the sediment grains, or clasts, that form the rock. Sandstones have medium-sized grains ranging from the size of beach sand to very fine, barely visible particles. Conglomerates, another kind of clastic sediment, have large grains—from very small pebbles the size of grains of rice to rocks bigger than a man's fist. Conglomerates are always very poorly sorted, i.e., there are a large variety of grain sizes in the rocks. Shale is yet another clastic rock; its grains are microscopic. In addition, shales contain various types of clay minerals.

Evaporites are formed when a saline body of water evaporates. As the salinity increases, certain chemicals precipitate and fall to the bottom, forming a layer. The most common evaporites are gypsum, anhydrite, and halite (rock salt). Halite is especially interesting to oilmen because many oil fields have been associated with salt domes, especially along the Gulf Coast.

Most organic sediment is classified as a carbonate, which includes limestone and dolomite. Carbonates occur in many forms: extremely fine-grained micrites from limestone mud, chalk from the excrement of golden-brown algae, and reefal limestones from coral. Organic sediments are usually the skeletons of small marine organisms that sink to the bottom and accumulate. Over time these organisms form layers that are often hundreds of feet thick. The layers are eventually buried, compacted, and cemented together, making a carbonate bed—an important reservoir rock.

All sedimentary rocks have some water in their pore spaces (the space between the individual grains). This water is referred to as formation water and contains varying amounts of dissolved salts. The source of the formation water may be the lake, river, or ocean in which the original sediments were deposited or, as in the case of sand dunes, the source may be rainwater or later groundwater which migrates in as the formations are buried.

All formations contain water in addition to whatever hydrocarbons they may contain. In Fig. 3–1 we see grains of various kinds of minerals forming a sandstone. These grains have been cemented together by heat, pressure, and a cementing material such as carbonate and/or silica. Since they were either laid down in a watery environment or were later submerged, the sandstone grains will normally be water wet, i.e., water clings to the individual grains. Nearly all formations have this immovable, or irreducible, water. No matter how much oil or gas a formation may contain, the irreducible water remains.

POROSITY

When sediments are deposited and compacted, they do not form a solid mass of rock. Spaces exist between the grains (intergranular pores). The amount of space, or voids, as a percentage of the total volume of formation is called the porosity (Fig. 3–1), or nonrock volume. Formation fluids (oil, gas, or water) accumulate in the voids. Without pores, a formation is of no interest to the oilman because there is no place for the rock to hold hydrocarbons. The larger the porosity, the more fluids a formation will contain.

Intergranular Porosity

To gain a clearer understanding of porosity, let's perform an experiment. Take a box that measures 1 ft on each side (volume = 1 cu ft) and fill it with uniformly sized marbles (Fig. 3–2a). The space between

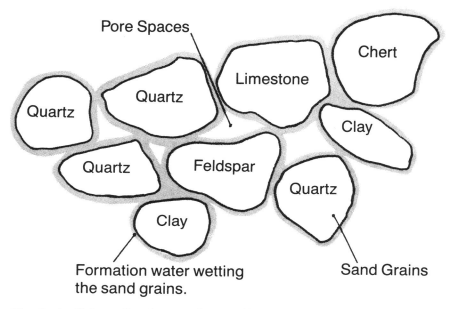

Fig. 3–1 Schematic of a sandstone. Note the formation water wetting the sand grains.

the marbles is the pore space. We can determine how large the pore space is by using a measuring cup to pour water into the box. If we measure very carefully, we will find that we can add 0.476 cu ft of water. This means the box of marbles has a pore space, or porosity, of 0.476 cu ft, or 47.6%.

The marbles in Fig. 3–2a represent sand grains. Obviously, sand grains are not symmetrical like marbles, so this experiment represents the best, or limit, conditions. Look at the way the marbles are arranged in the box, stacked on top of each other. Sand grains seldom align themselves like this in nature because this arrangement is unstable. Here we have the largest possible porosity; so if we see a porosity reading greater than 47.6% on a log, we know we are not looking at a sandstone formation.

An odd but interesting mathematical fact is that the porosity will not change with the size of the marbles. Whether we put a 1-ft-wide marble in the box or carefully stack it full of 1-mm marbles, the porosity will be the same as long as the arrangement is the same. If you're good at simple geometry, you can prove this to yourself. The equation for the volume of a sphere is:

$$V = \tfrac{1}{6} \pi d^3$$

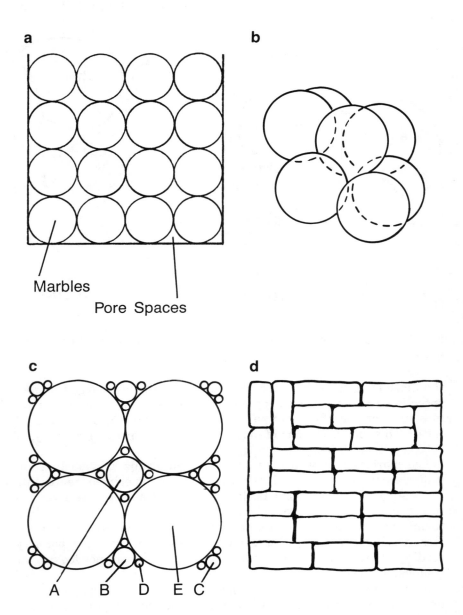

Fig. 3–2 Porosity and how it changes relative to grain shape and arrangement. (a) Maximum possible porosity; porosity ≈ 47%. (b) Rhombohedral stacking; porosity ≈ 26%. (c) Various-sized grains (A,B,C,D,E) in a closely packed arrangement; porosity ≈ 10%. (d) Loosely stacked bricks; porosity ≈ 1%.

Where:

V = volume
d = diameter of the sphere
π = 3.1416

What happens if we alter the arrangement of the marbles? In Fig. 3–2b the marbles are arranged in a rhombohedron, that is, they are stacked together the way you'd stack cannon balls. The maximum possible porosity now is 26%. This arrangement is stable and is more likely to exist in nature. We can say that a well-compacted, well-sorted (sand grains are essentially the same size) sandstone should have a maximum porosity of around 26%.

If we vary the size of the marbles, what happens? Look at Fig. 3–2c. Four marbles are stacked on top of each other; we know their porosity is 47.6%. If we add smaller marbles that will fit in the spaces between the big marbles, we will reduce the original porosity. How much we reduce it depends on how many different sizes of smaller marbles we use. If we use only size A marbles, we reduce the porosity very little because there is room for only one size, A—right in the middle of the arrangement. If we use both A and B sizes, we can add four Bs and reduce the pore space more. If we use size A, B, C, D, and E marbles, we can practically eliminate the porosity. Note that we get the greatest reduction in porosity with a wide assortment of marble sizes. From this we can say that, in general, poorly sorted sandstones have low porosities.

What happens if we replace the round marbles with bricks (Fig. 3–2d)? The only pore spaces are in the cracks between the stacked bricks, so the porosity is very, very low. We can learn from the brick illustration what kind of porosity we get with square grains. But first let's digress a bit.

If we chip off all the sharp edges from a cube, eventually we end up with a sphere. Nature, too, usually begins by breaking off a more-or-less rectangular piece of rock and then proceeds to turn it into a sphere or a marble by rolling, grinding, and wearing off the sharp edges. A freshly broken piece of rock is angular; a very weathered fragment is well rounded. Between these stages the rock passes through subangular and subrounded stages.

The more angular the sand grains, the more tightly packed they will be and the lower the porosity. The more rounded the grains, the higher the porosity.

Usually well-rounded sands are also well sorted, so they have very high porosities and permeabilities. (Permeability is the ability of a formation to allow fluid to flow. High permeability results in high production, low permeability in low production—other things being equal.) These sands are described as mature, well-weathered, or even beach sand (if the grains are large). On the other hand, poorly sorted, angular, and subangular sands are said to be immature and slightly weathered with low porosities and permeabilities.

Other Types of Porosity

A second lesson Fig. 3–2d illustrates is fracture porosity. If the bricks are rather loosely stacked, there will be spaces between them. These spaces are more in the nature of fracture planes than pore spaces. The total pore volume of a fracture system will usually be very low, often 1–2%. Fractures occur naturally in rocks because of the way the earth moves and buckles over time. Although fractures have a very low porosity, they frequently have a very high permeability; large quantities of fluid can flow very easily.

Another type of porosity is vugular porosity, present in carbonates such as limestones and dolomites. A vug is a large, irregular void in the rock, usually caused when a mineral, such as calcite, is dissolved by water moving through the rock. Vugs allow large quantities of fluid to flow very easily. Caverns are examples of huge vugs.

A particular formation could have all three porosity systems or only one. Sandstones normally have only the first type of porosity, called matrix or intergranular porosity. Carbonates often have all three porosity systems: matrix, fracture, and vugular.

FORMATION ANALYSIS

Fig. 3–3 illustrates some important concepts in formation analysis. Part (a) depicts a unit volume of formation. This unit volume measures one unit (I) per side and has a volume of 1, or 100%. The sand grains form the matrix or structure of the rock. The formation water wets the sand grains, and the oil is in the pore space not occupied by the water. This particular unit volume has only irreducible water and oil in the pore space. Note that the pore space is 100% full of fluid, either oil or water.

Now look at Fig. 3–3b. Here the unit volume has been separated into its constituents. All of the sand grains have been compacted into the bottom and are called the matrix, the mineral from which the rock structure is made. In the case of a sandstone, the matrix is predominantly quartz.

Since we have a unit volume and all of the compacted matrix is at the bottom, whatever is left over is the porosity. The pore space is filled

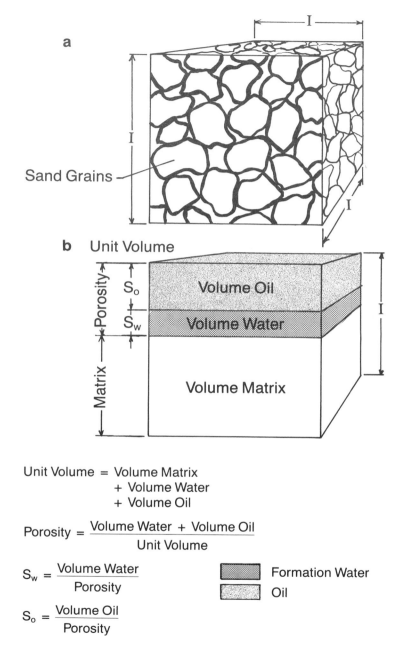

Unit Volume = Volume Matrix
    + Volume Water
    + Volume Oil

$$\text{Porosity} = \frac{\text{Volume Water} + \text{Volume Oil}}{\text{Unit Volume}}$$

$$S_w = \frac{\text{Volume Water}}{\text{Porosity}}$$

$$S_o = \frac{\text{Volume Oil}}{\text{Porosity}}$$

Fig. 3–3 Unit volume and bulk volume. (a) Example of how fluids and pores are distributed in a formation. (b) The same formation and fluids separated for easy analysis.

with water and oil, so they are shown in the area labeled "porosity."
If we add up volumes, we can say:

$$\underbrace{\text{BVM}}_{\text{Volume matrix}} + \underbrace{\text{BVW}}_{\text{Volume water}} + \underbrace{\text{BVO}}_{\text{Volume oil}} = 100\% \text{ or } 1.0$$

These volumes are called bulk volumes. In the figure, the bulk volume matrix (BVM) is 0.70 (70%); the bulk volume water (BVW) is 0.10 (10%); and the bulk volume oil (BVO) is 0.20 (20%).

Now we know that all of the oil and water is contained in the pore spaces; so if we add BVW and BVO, we find the porosity, 30% or 0.3. We can say:

$$\text{Porosity} = \text{BVW} + \text{BVO} = 100\% - \text{BVM}$$

Although the concept of bulk volume is very useful, traditionally we have used water saturation, S_w, and oil saturation, S_o, to account for the liquids that occupy the pore space. To determine these saturations, we calculate the following:

$$S_w = \text{BVW}/\text{Porosity}$$
$$S_o = \text{BVO}/\text{Porosity}$$

Notice that $S_w + S_o = 1$ or 100%. This means the pore space is 100% full.

Note: If free gas is present, its saturation is noted by S_g. In a formation with oil, gas, and water $S_o + S_g + S_w = 100\%$.

Shaly Formations

So far we've talked about shale-free or clean formations. But quite often the formations we encounter contain varying amounts of shale, which complicates things immensely.

Shales and clays (the terms are used almost interchangeably although there are technical differences) are very fine-grained, plate-like minerals. Because of their platy structure and small grain size, they have immense surface areas compared to the same volume of sand grains. The effect is to bind large quantities of water to their structures. Because of their fine grain size and the strong forces that hold the water in place, shales have essentially zero permeability and, usually, high porosity.

The shale can occur as fine laminations in the sand, or it can be dispersed throughout the formation. If dispersed, the shale acts almost like another liquid in the pore space: it lowers the porosity that is available to hold fluids yet reduces the permeability. The porosity that is available to hold fluids is called the effective porosity, as opposed to

Formation Parameters   33

the total porosity. The total porosity includes the porosity of the shale which is filled with bound water.

We now have a revised unit volume (Fig. 3–4): the shale matrix is included with the formation matrix (the shale and sandstone matrices are very similar except for grain size) and the shale porosity is included in the porosity section. We can determine the effective porosity by subtracting the shale bound water volume from the total porosity. The effective water saturation is then determined by dividing the bulk volume of the formation water by the effective porosity.

Reserves Estimate

From Fig. 3–4 we see that we can calculate the amount of oil or gas in place in the formation if we know the effective porosity, the water saturation, the formation thickness, and the area that the reservoir covers or that the well is capable of draining. To calculate the reserves (the amount of recoverable oil or gas), all we need to know in addition is a recovery factor (oil recovery factors are usually around 40% but may be much higher or lower). We arrive at the following equation:

Oil Reserves (N_p) = Effective Porosity (ϕ_e) × Oil Saturation (S_o)
   × Formation Thickness (h) × Drainage Area (A)
   × Recovery Factor (rf)

From the above equation we can see that we can recover more oil if:

- Effective porosity is higher
- Oil saturation is higher (water saturation is lower)
- Formation is thicker
- Drainage area is larger (bigger reservoir)
- Recovery factor is higher

We can only estimate some of these quantities, such as recovery factor (use a best-case/worst-case scenario) or drainage area (estimated from spacing requirements, seismic information, leased area, or a best guess). The other quantities—porosity, oil saturation, and thickness— are measured by the logs. With a reserve number we can perform economic evaluations to determine if the well will pay out.

INVASION

So far we have been looking at the reservoir in its undisturbed state. However, drilling can profoundly affect the characteristics of a for-

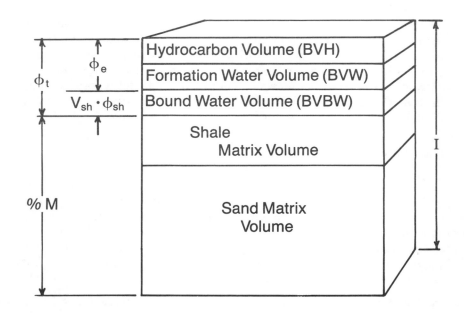

Unit Volume = 100% = BVM + BVBW + BVW + BVH
= Volume Sand Matrix + Volume Shale Matrix + Volume Bound Water + Volume Formation Water + Volume Hydrocarbon

Total Porosity = ϕ_t = $\dfrac{BVBW + BVW + BVH}{Unit\ Volume}$

Effective Porosity = ϕ_e = $\phi_t - V_{sh}\ \phi_{sh}$

$S_{w_e} = \dfrac{BVW}{\phi_e}$

$S_{w_t} = \dfrac{BVW}{\phi_t}$

Note: BV = Bulk Volume

Fig. 3–4 Unit volume and bulk volume, shale added. Note how the addition of shale changes the relationships from Fig. 3–3.

mation. The drill bit changes the rock somewhat, but the main alterations are caused by the drilling mud.

Drilling mud is a complex liquid usually composed mainly of water and suspended solids (weighting materials) as well as various chemicals that control the mud's properties (viscosity, fluid loss, acidity). Mud carries the cuttings out of the hole and up to the surface. Clays are added to water to give body to the drilling fluid. This combination makes a better carrying agent than plain water.

Another important use of the mud is to control formation pressure. Weighting materials such as barite are added to the mud so that the hydrostatic, or fluid, pressure of the mud column is greater than the formation pressure. This excess pressure stops the well from flowing or "kicking" during drilling operations. If a high-pressure zone is encountered during drilling and if formation pressure exceeds hydrostatic pressure, the driller must weight up (add weighting materials) until the well is under control and pressure is balanced again.

If we take a sample of the mud and place it in a mud press, we can separate the mud into its two main components: mud filtrate and mudcake. Mud filtrate is a clear fluid whose salinity varies according to the source of the drilling water (the water used to make the mud) and the additives. Usually the filtrate salinity is lower than the formation water salinity. Since the filtrate is a clear fluid (no suspended solids), it can invade the formation if the pressure in the wellbore is greater than the formation pressure and can displace some of the original fluids.

The solid component of the mud is called the mudcake. The mudcake seals off the formation from invasion by the mud filtrate. The presence of mudcake can be detected by some logging tools. It is an indication of invasion and, indirectly, of permeability. Since the mudcake is a solid, it will not normally invade the formation. Drilling mud usually cannot invade the formation because it contains a lot of suspended solids. Whole mud can be lost into the formation by inadvertently fracturing the formation or by natural fractures or vugs. In the case of fracturing, the formation strength is exceeded by the hydrostatic mud pressure, and a fracture large enough to take the mud solids is formed. This is commonly called lost circulation or a lost circulation zone, and large amounts of mud can be lost in a short time.

We've separated the mud into two components, filtrate and mudcake. Let's see what happens when these two components come in contact with the formation.

On the left side of Fig. 3–5 is a portion of undisturbed formation. The formation we are interested in is the section of sandstone bounded above and below by shales. Note that the bottom of the sand has a

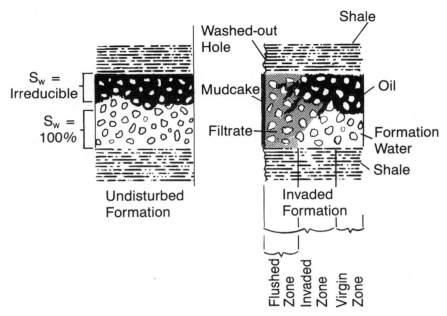

Fig. 3–5 Invasion. Note differences between undisturbed formation (left) and invaded formation (right).

water saturation, S_w, of 100% and the upper part of the sand is at irreducible water saturation, $S_{w_{irr}}$. (Only the irreducible water remains; oil fills the rest of the pores.) The water fills the lower portion of the formation because oil is less dense or lighter than water and rises. A transition zone exists between the upper and lower sections in which the water saturation is changing from 100% to irreducible. (Note that not all zones containing hydrocarbons will be at $S_{w_{irr}}$.)

On the right side of Fig. 3–5 is the same formation after it has been penetrated by the bit. Here, invasion has occurred. Because of the higher hydrostatic pressure of the mud column, the formation acts like a mud press and separates the mud into mudcake, which is plated out on the borehole wall (note that the hole diameter is less than bit size where we have mudcake), and filtrate, which invades the formation. In order to invade the formation, the fluids that were originally there must be displaced. The filtrate flushes or displaces these fluids deeper into the reservoir and takes their place near the wellbore.

In the bottom part of the formation where $S_w = 100\%$, the flushing is nearly complete. Most of the formation water can be moved because it is being displaced by water that is different only in the amount of dissolved salts. (In fact, sometimes the filtrate and the formation water

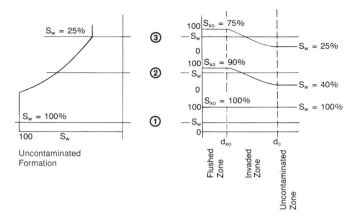

Fig. 3–6 Change in water saturation for Fig. 3–5. Saturation varies with distance from the borehole.

are nearly the same salinity. In that case it is impossible to tell whether the formation was invaded.) The salinity of any formation water left behind will soon reach equilibrium with the filtrate because of ion exchange.

In the upper part of the zone, we have a large oil saturation ($S_w = S_{w_{irr}}$). Although most of the formation water has been displaced by the mud filtrate, some oil, called residual oil, still remains in the flushed zone.

Residual oil is similar to irreducible water saturation: it cannot be moved by normal means. The residual oil saturation may be abbreviated S_{or} or ROS. The residual gas saturation is designated S_{gr}. The term S_{hr}, residual hydrocarbon, may be used for either gas or oil. If a formation has never contained any hydrocarbons, it will have an S_{hr} of 0. Once the formation has contained oil or gas, even if it was only migrating through the formation (that is, once S_w has been less than 100%), the formation will have an S_{hr} greater than 0.

Look again at Fig. 3–5. The filtrate has flushed out all the original fluids possible as far as a certain depth; this depth is called the flushed zone. The subscript for the flushed zone is "xo," so the flushed zone water saturation is S_{xo}. The diameter of the flushed zone is d_{xo}, the resistivity of the water in the flushed zone is R_{mf} (resistivity of mud filtrate), and the resistivity of the flushed zone formation is R_{xo}. If we go a little deeper into the formation, we will find a mixture of formation fluids and mud filtrate. This zone, from the borehole wall to the end of the mud filtrate, is called the invaded zone and has a subscript of "i." The diameter of invasion is called d_i; the water saturation is S_i;

and the water has a resistivity of R_i. Since the water in the invaded zone is a mixture of formation water and filtrate, it is usually impossible to come up with a value for R_i. Finally, as we pass the invaded zone, we return to the undisturbed or uncontaminated formation. This is the virgin zone; here, the conditions are the same as on the left side of the figure.

Fig. 3–6 is the same formation presented in a different fashion. On the left is a plot of S_w vs. vertical depth. At the bottom, $S_w = 100\%$ and is constant for 30 ft or so. We then enter the transition zone, where S_w changes with depth until it reaches an irreducible value of about 25%. S_w is constant for the last 20 ft at $S_w = S_{w_{irr}} = 25\%$. Note that S_h, the hydrocarbon saturation, is equal to $1 - S_w$.

On the right side of Fig. 3–6 are three sections, drawn horizontally through the formation so that we can see how S_w varies with distance from the borehole at three different points in the formation. Look at the bottom section first. This represents the $S_w = 100\%$ formation. Note that $S_w = 100\%$ throughout, i.e., $S_{xo} = S_i = S_w = 100$. This is because there was never any oil or gas in this part of the formation. Since we haven't added any hydrocarbons, all of the water saturations must read 100% and S_h must read zero.

Now look at the middle section. This was taken in the transition zone, where the S_w was about 40%. Now we see a change in the various water saturations because oil is present in the formation and some of it has been flushed out by invading drilling fluids. Because of the residual oil, S_{xo} will be less than 100%. S_i will be lower than S_{xo}, and of course S_w in the uncontaminated zone will be 40%.

In the upper section, S_w is at its irreducible saturation value. Here we see a maximum variation in the various saturations. S_{xo} will be lower than the S_{xo} in the transition zone, and $(1 - S_{xo})$ will be close to the residual oil saturation, S_{or}. The water saturation will vary throughout the invaded zone between d_{xo} and d_i. Beyond d_i, $S_w = S_{w_{irr}} = 25\%$. In this upper section we can clearly see the flushed-zone diameter, where S_{xo} is constant, and the end of the invaded zone, where S_w becomes constant.

Be sure to study Figs. 3–5 and 3–6 thoroughly. Many of the dilemmas in interpretation and evaluation arise from uncertainties about invasion.

DIRECT VS. INDIRECT MEASUREMENTS

Before we discuss resistivity, let's talk about measurements. We need to know many things during the day, and most of these can be measured: our weight, our shoe size, the time, the number of gallons of gasoline we pump into our cars. We almost take for granted our ability to make these measurements.

We assume that when the butcher weighs a pound of meat, we are getting a true pound. But how do we know that? Where does the butcher keep his pounds? Actually, he is measuring weight indirectly. He places the package of meat on the scale platform and then notes how far a spring stretches. The stretch of the spring is proportional to the weight of the object. If the scale has been calibrated, he can tell how much the object weighs in some unit, such as pounds or grams. But he does not weigh the meat directly; he weighs it indirectly by stretching a spring.

The same is true in engineering measurements. Seldom is it possible or practical to measure something directly, so indirect methods must be used.

As we pointed out earlier, to calculate reserves we need to measure (1) porosity; (2) the percentage of formation fluids that are water (water saturation, S_w) and gas/oil (gas saturation, S_g, and oil saturation, S_o); (3) formation thickness (h); (4) recovery factor; and (5) area of the reservoir. Some of this data, such as formation thickness, can be obtained easily from logs; other items, such as reservoir area, must be estimated by various means such as seismic data and offset wells. Recovery factor is another item that can only be estimated initially. Log data indicate only whether hydrocarbons will flow from a given formation; years of data from a particular reservoir are needed to determine the ultimate recovery factor. Log evaluation therefore concerns itself primarily with determining porosity and water saturation. From this information we can infer production potential.

Porosity is measured in the laboratory in a manner similar to the experiment with the marbles. Unfortunately, porosity can not be measured in the same manner in situ (in place in the wellbore). Therefore, various indirect methods must be used. Most of these methods use either sonic energy (the response of the formation to a sound wave passing through it) or some form of induced or applied radiation.

Neither water nor hydrocarbon saturations can be measured directly in the wellbore. However, it is possible to infer the water saturation fairly easily, if the porosity is known, by measuring the resistivity of the formation. In fact, this was the earliest wireline log: the resistivity log.

RESISTIVITY

What is resistivity and how do we use it to determine water saturation? Resistivity is a term that expresses the difficulty an electric current has in flowing through a substance. For example, copper has low resistivity; electric currents meet little resistance in copper objects such as power lines and flow easily. On the other hand, glass has very high resistivity and is often used to make insulators. Conductivity is the opposite of resistivity, so copper has very high conductivity while glass has very low conductivity. The resistivities/conductivities found in the earth's formations fall somewhere between those of copper and glass.

Since sandstones are essentially glass (silicon oxide), it would seem that sandstones would have very high resistivities. That is true to a certain extent. If you have a block of completely dry sandstone and measure its resistivity, the reading will be high—approaching infinity. But we know from looking at logs that sandstones often have resistivities of less than 1 ohm, a very low value. (Although we speak of resistivity values in ohms, this is an abbreviation for "ohmmeters squared per meter." An ohm is a measure of resistance, just as pounds are a measure of weight.) So how do we resolve the differences in the resistivity between a dry sandstone and one that we find downhole? The difference is the formation water, which usually has a low resistivity.

If we measure the resistivity of distilled water, we will find it has an extremely high resistivity. So why does formation water usually have a low resistivity? Could it be the dissolved salts? Yes. The dissolved salts that are generally found in formation water in varying amounts lower the resistivity in the formation water. This means that when we measure the resistivity of our block of sandstone, its resistivity will vary with the amount of water and the salinity of the water.

Now let's set up another experiment. We'll devise an apparatus to measure the resistivity of a block of formation, and we'll call it an R-meter (Fig. 3–7). Start off with a piece of completely dry sandstone—nothing in it except sand grains and air—and measure its resistivity. The resistivity should be close to infinite. Now take some distilled water, saturate the dry sandstone block with it, and measure the resistivity again. Once more it will be close to infinity. Finally, make up a saltwater solution, measure its resistivity, R_w, then flush out all of the fresh water with the brine. If we measure the resistivity of the block now, we'll see that it is much lower than before. Record these readings: R_w, the saltwater resistivity, and R_o, the resistivity of the water-saturated block ($S_w = 100\%$).

Now redo the experiment with a different R_w; this will give us a different R_o. If we repeat the experiment many times, we will see a relationship between R_w and R_o. This relationship is called the formation resistivity factor, F_r; it is expressed mathematically as $F_r = R_o/R_w$.

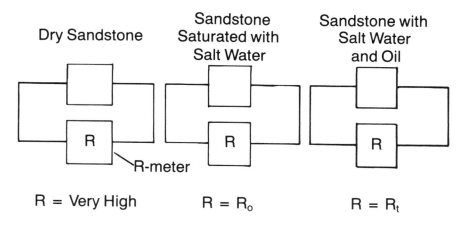

Fig. 3–7 Hypothetical R-meter experiment. By comparing different values of water resistivity (fresh and saline), we can deduce the formation resistivity factor, F_R.

If you think about what we're doing when we determine F_r, you'll conclude that the formation resistivity factor is related to porosity in some way. Not only the salinity but also the amount of water in the rock must have a bearing on R_o. What affects the amount of water that the rock can hold? The relationship is written $F_r = K_R/\phi^m$, where K_r is a constant, usually between 0.8 and 1, and the exponent m is usually between 1.3 and 2.5.

If we change our experiment and add some oil to the block, changing S_w, we can come up with a new relationship. First measure R_o by

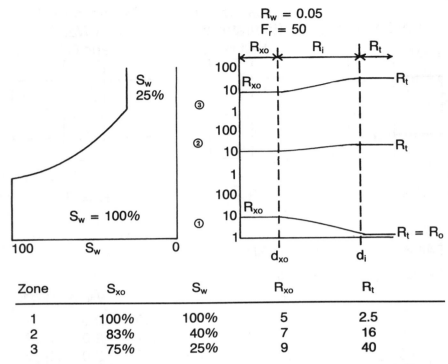

Fig. 3–8 Resistivity profile. Figure indicates invasion by mud filtrate. Compare with the profile in Fig. 3–6.

saturating the rock with saltwater. Next reduce S_w to 80% by adding the oil and measure the resistivity of the rock. We will call this reading R_t for true resistivity. Adding oil will increase the resistivity of the block so that R_t is greater than R_o. Now if we reduce S_w again to, say, 60%, measure R_t again, then reduce S_w again and again for many samples of different kinds of rocks, we will come up with the saturation equation, also called the Archie equation.

These are essentially the experiments that George Archie, who pioneered studies in resistivity, carried out in the early days of log interpretation research. The "father of log interpretation" ran experiments in which he first measured the resistivity of a 100% water-saturated core and then measured the resistivity as the core was progressively saturated with oil. Archie determined that water saturation is equal to the square root of the 100% water-wet resistivity, R_o, divided by the formation resistivity, R_t:

$$S_w = \sqrt{R_o/R_t}$$

Formation Parameters   43

Now that we've talked about resistivity, let's take one last look (for now) at invasion and see what the resistivities would look like in different parts of the formation.

The presence of invasion by the mud filtrate gives rise to a resistivity profile, as shown in Fig. 3–8. Here the water saturation varies from 100% in the bottom of the zone through a transition zone to $S_w = 25\%$ at the top. Notice the three resistivity profiles on the right side of the figure. On all of the profiles the flushed zone resistivity, R_{xo}, next to the borehole is very high. It is slightly higher in the oil-bearing zones, because of the residual hydrocarbons, than in the 100% wet zone.

In the uninvaded or virgin part of the formation, there is a large contrast in resistivity to the flushed zone, where $S_w = 100\%$. The contrast is not nearly as great through the transition zone, and there is very little contrast in resistivities in the low water saturation interval.

The resistivity close to the wellbore is generally higher than the true resistivity of the formation, even when the formation has hydrocarbons. (It is possible to have R_{xo} less than R_t when drilling with very low-resistivity mud systems or when the water saturation is very low.) Since the depth of invasion is unknown, tools have been developed that measure the resistivities at various depths around the wellbore. We will study these tools in Chapter 5. Before we advance to that topic, though, we need to study one of the most basic logs of all: the mud log.

4

MUD LOGGING

As the drill bit cuts through the different formations, the cuttings are brought to the surface by the drilling mud. Traces of oil or gas may also be brought up in the mud. The practice of mud logging tries to identify, record, and/or evaluate lithology, drilling parameters, and hydrocarbon shows. The information obtained by the mud logger is presented in the form of various logs such as the driller's log, the cuttings log, or the show evaluation log. The mud logger takes this information, correlates it with data from other wells, and determines whether the well may be able to produce hydrocarbons in commercial quantities. In addition, the mud logger monitors the wellbore for stability to prevent blowouts or kicks, and he makes sure information is relayed to the right people at the right time.

To comprehend all of the information available, we need to understand four important areas of mud logging: rate of penetration and lag, gas detection, formation evaluation and sample collection, and show evaluation.

RATE OF PENETRATION AND LAG

Rate of penetration (ROP) is the oldest and most common way of measuring and evaluating formation characteristics and drilling efficiency. The formation's lithology (rock type and hardness), porosity, and pressure affect the ROP. The drilling parameters that affect ROP include the weight on the bit, the bit's speed (rpm), the drillstring configuration, the type of bit selected and its condition, and hydraulics.

Measuring ROP

Mud loggers measure ROP manually in three main ways: strapping the kelly, observing the drilling rate curve, or checking Geolograph charts. When the rig crew straps the kelly, they mark it in some in-

---

Much of the material from this chapter was summarized with the permission of Anadrill, a division of Schlumberger, from their Delta Training Manual, Vol. 1.

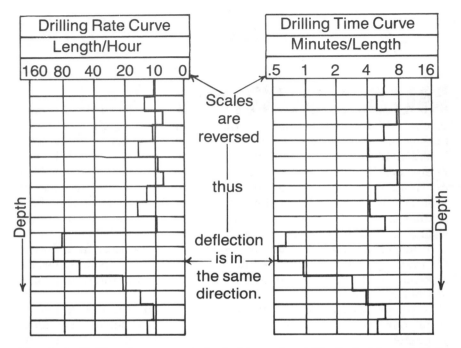

Fig. 4–1 Drilling time curve (adapted from Anadrill's *Delta Manual,* 4-4). Note that units may be either in length per hour or in minutes per length; yet the curves are essentially identical.

crement of depth (feet or meters) and then record how much time is needed to drill each interval. When using the drilling rate curve (Fig. 4–1), the logger observes a circular or strip chart, graduated in units of time, which moves a pen calibrated in units of depth. He notes the time taken to drill a certain number of feet or meters and can then determine ROP. The Geolograph chart, a third means of measuring ROP, is a strip chart on a drum that rotates once every 24 hr. Like the drilling rate curve chart, this chart is also marked in intervals of time. Here, though, each increment of length is recorded as a tick mark.

When rate of penetration is presented, it is a plot of ROP vs. depth. Usually it is included on mud logs along with the parameters for correlation and interpretation. ROP can also be plotted automatically with an on-line plotter.

If ROP is expressed in units of length/hr, its curve is called a drilling rate curve. When the units are in min/length, the curve is referred to as a drilling time curve.

Mud Logging 47

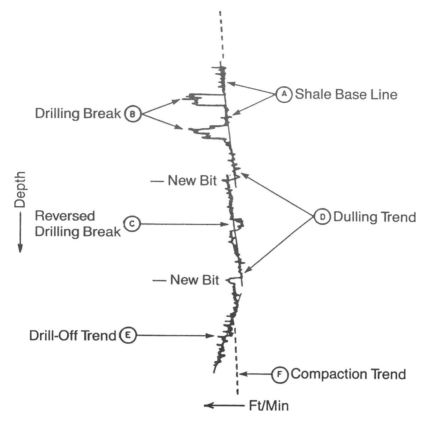

Fig. 4–2 ROP curve terminology. (courtesy Anadrill, *Delta Manual,* 4-7)

Interpreting ROP from the Mud Log

Fig. 4–2 is a simplified ROP curve. On it are several lettered areas which illustrate the kinds of information we can read from the log.

The base line (A) is a reference point for interpretations; it simplifies correlations. Usually the rock chosen for the base line is one of the hardest lithologies that the bit will drill through. In this formation, we use shale as the base line; however, limestone is often used as the base line in carbonate sequences. Whatever the lithology, the base line establishes a norm.

Any deflections from the norm signal changes in formation. In this example, the deflections are interpreted as a sand/shale sequence. The drilling break (B) is the deflection. It usually indicates a change in lithology, although it sometimes is due to crossing a fault. Whatever the case, the drilling break notes an abrupt increase in the ROP—usually two or more times greater than the base-line average.

Occasionally we note a reversed drilling break (C) on the track. This indicates an abrupt decrease in the rate of penetration and can imply changes in lithology. Reverse drilling breaks usually are associated with very dense formations called "caps." They may also denote a shale/sand interface or a formation where production has depleted the formation's pressure.

As the bit wears out, it drills less efficiently. The ROP curve shows this change as a slope away from the base line (D). The dulling trend can help the driller know when a bit needs to be changed.

A drill-off trend (E) is a gradual, usually uniform increase in the ROP. It often indicates a transition zone where pore pressures are increasing.

As the overburden pressure and age of the rocks increase with depth, the formation becomes more compacted. This compaction trend (F) can sometimes be seen on the log over long intervals. See Fig. 4–3 for the drilling responses of some common rock types.

Lag

Lag is the amount of time that elapses from the moment when the bit penetrates a new formation until the moment when the downhole particles and/or traces of gas travel back up the wellbore to the surface. Rate of penetration is measured instantly; as soon as the bit increases or decreases its speed, the driller has input. However, samples from that particular formation may not be circulated up for several minutes. The mud logger must keep this in mind as he correlates the data.

GAS DETECTION

Gases extracted from the mud system are usually the first indication that hydrocarbons are present downhole. Gas enters the drilling fluid from one of three sources: (1) a gas-bearing formation, (2) a formation feeding gas into the mud, or (3) contamination.

As the bit drills through a formation, it opens or exposes some of the pores. Fluid from these opened pores mixes with the drilling mud. This gas, along with cuttings from the destroyed formation, is pumped back up toward the surface. As the gas and cuttings rise, the pressure drops

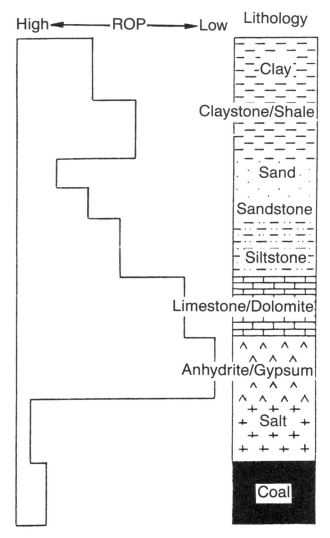

Fig. 4–3 Drilling responses of common rock types. (courtesy Anadrill, *Delta Manual*, 9-3)

and more gas comes out of the pores in the cuttings. This "liberated gas" is an important piece of data for log interpretations.

If the hydrostatic pressure is less than the formation pressure, even more gas can flow into the wellbore. The amount of flow from the formation into the borehole depends upon the pressure differential (the difference between the hydrostatic pressure and the formation pressure), the porosity and permeability, the properties of the formation's fluids, and the length of time this condition lasts. When the formation fluids enter continuously, the well is said to kick. Swabbing (lifting the drillstring rapidly) also encourages formation fluids to flow into the well because the wellbore pressure drops. Engineers can identify this gas that enters the borehole during swabbing, called connection gas, and can use the data to enhance formation evaluation and improve well safety.

Occasionally gas is introduced into the drilling fluid from a source other than the formation, particularly when oil-based drilling fluids are used. This is called contamination gas. However, this case is rare.

GAS DETECTORS

Gas in mud is measured when the drilling fluid returns to the surface. A collector or trap is used to remove gas samples from the return line. The trap samples the mud returns consistently and reliably, regardless of the circulation system's flow rate. In addition to continuous sampling, random batch samples are collected and the gas components are extracted.

Gas can be detected in five main ways: thermal catalytic combustion (TCC) or hot-wire detector (HWD), gas chromatography (GC), thermal conductivity detector (TCD), flame ionization detector (FID), or infrared analyzer (IRA).

TCC instruments, more commonly known as hot-wire detectors (HWD), have been around for a long time because they are simple and inexpensive, and they perform adequately. However, the instruments are unstable so responses vary, and the method fails at gas concentrations that exceed a few percent. Nevertheless, this technique remains primary, especially when supplemented by other procedures such as gas chromatography.

Gas chromatography results are more accurate and more quantitative than hot-wire methods; however, they take minutes rather than seconds to complete. Therefore, TCC is used to detect the presence of hydrocarbons in the mud returns, while chromatography is used to analyze the composition of the gas stream on a regular but intermittent basis, usually after a show is detected with the TCC.

The *thermal conductivity detector* is the least sensitive device normally used for monitoring hydrocarbons. Under optimum conditions, the detection limit of hydrocarbon in air is about 1%. However, the TCD has good linearity, is easy to use, and is durable and inexpensive.

Flame ionization detectors are popular for gas analysis instrumentation used outside of the mud logging industry. They are superior in many ways to other systems. However, the FID is expensive and difficult to operate, which limits its use.

The final device, *the infrared analyzer,* may be operated continuously, but only for one compound at a time. In addition, IRAs cost more than hot-wire devices and are sensitive to vibration and power supply variations. Otherwise they are easy to operate and have a sensitivity comparable to TCC methods.

A combination of two or more of these five methods helps mud loggers detect the presence of gas and analyze its components.

ANALYZING RETURNS

The entire detection and analysis process is very orderly. The sequence yields data on the content of gas in mud, oil in mud, gas in cuttings, and oil in cuttings. These data constitute the hydrocarbon log, a continuous record organized by well depth.

The continuous gas sample is usually analyzed by the TCC or HWD gas detector. These devices are not calibrated to an absolute scale, so any gas responses or "shows" are relative. The results are usually reported in "units," such as a 200-unit show.

After a show is detected with the TCC, a sample is analyzed with the gas chromatograph. This device reports the analysis of the gas as percentages of methane (C_1), ethane (C_2), propane (C_3), butanes (C_4), and pentanes (C_5). In gas associated with an oil show, there will be a higher percentage of C_3, C_4, and C_5. In a gas show, C_1 and C_2 will be predominant.

If the gas saturation of the oil is low, an oil show may not be detected by the gas-in-mud response. In these cases, loggers try to detect the presence of oil in the mud in other ways. Sometimes oil can be identified simply by visual examination. Since the mud is largely water, the oil will float on the surface of the water and may be detected by color sheen or oil globules on the surface of the mud pit or the samples. Color, intensity, and fluorescence (oil fluoresces under a black light) are then noted.

Gas may also be detected in fresh cuttings. A sample of cuttings is pulverized in water within a sealed container. After pulverization, the vapor above the cuttings-fluid level is analyzed with a hot-wire detector or gas chromatograph. The logger also looks at the sample to detect any shows of oil, such as sheen or fluorescence.

Measuring and Recording the Readings

Total mud gas readings are recorded continuously with a recorder chart. The gas readings are recorded for each logging interval and represent the gas response vs. depth. Since there is a lag between the time when a formation is drilled and the time when the sample reaches the surface, gas readings must be lagged.

In logging total gas readings, several measurements are usually designated separately:

Total Gas Reading—The maximum reading during a specific interval or the total meter reading at any point.

Background Gas Reading—Either drilling background gas, which is the average gas reading while drilling in low-permeability zones such as shale, or circulating background gas, the average reading when circulating the bit off bottom.

Connection Gas Reading—The difference between the drilling background gas and the total gas reading that occurs during a drillpipe connection.

Trip Gas Reading—The maximum total gas reading from bottoms up after a trip.

A data pad (Fig. 4–4) is used for recording gas data. Several kinds of data can be recorded.

Columns A through G on Fig. 4–4 are fairly common for all forms. Column A records when a connection is made (a new section of drillpipe is added), and column B records the interval drilled. In column C, the drilling rate is determined for each interval.

The stroke counter counts the number of strokes made by the reciprocating mud pump. The end strokes reading, column D, records the end of each interval from the stroke counter. A lag stroke counter, which is set a certain number of strokes (the lag) behind the stroke counter, is also used. When the lag stroke counter equals the end stroke counter, the interval just drilled will be at the surface.

Column E, the total mud gas reading, records the reading for each depth interval when the lag counter reading equals the end stroke reading in column D.

Gas peaks that occur from connections, trips (when the entire drillstring is pulled from the hole, as in changing a bit), surveys (measurements such as hole deviation taken on the drillpipe), and other down (nondrilling) time should be designated in column F. Let's say we make a connection at 8,043 ft. We observe an abnormal gas peak of 90 units during the lagged interval 8,040 to 8,045 (the time required

Mud Logging 53

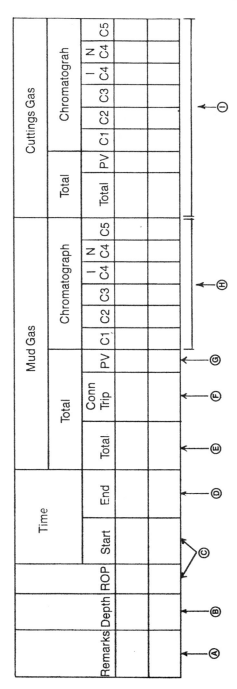

Fig. 4–4 Sample data pad (adapted from Anadrill's *Delta Manual*, 7-16). Note the discussion in the text for the values of each column.

to circulate cuttings from the bit to the surface). This gas peak is due to the connection gas, gas which seeps out from the formation during down time. The value of the connection gas is the total gas reading minus the background gas.

Column G is for reading the vapors of the thermal catalytic combustion filament. The value represents the presence of heavier hydrocarbons in the sample. In some areas, this type of detection is required in addition to chromatography, which is recorded in column H.

The final columns are used in areas where the cuttings are collected, broken up, and measured. The gas content readings are called "cuttings gas readings" and are used primarily to interpret permeability.

The total gas readings from column E are presented on the recorder chart as a continuous curve that records data immediately. They are also plotted on the formation analysis log, the show analysis log, and/or the pressure evaluation log. The most common correlation curve is the ROP curve; it is usually presented on any log on linear or nonlinear scales with bar graphs or point-to-point plots (Fig. 4–5).

Total gas measurement can be applied in three ways. First, it evaluates hydrocarbon shows. If the reading increases, hydrocarbons are present in a zone. The reading itself does not indicate productivity; however, increased readings in a potentially permeable, high-porosity zone often indicate the zone may be productive. Second, gas measurement detects pressure. Increasing background gas usually indicates increasing formation pressures. Connection gas and abnormally high trip gas usually indicate a nearly balanced mud system (hydrostatic pressure = formation pressure) or even an underbalanced system. Finally, the total gas reading curve can be correlated with other measurements, such as resistivity, ROP, spontaneous potential, and offset well curves. These correlations can provide input on potentially productive zones.

Interpretations

When the drill bit penetrates a hydrocarbon-bearing formation, only a small amount of hydrocarbon enters the mud and is mixed with large quantities of circulating fluid. The size of shows depends on drilling and sampling factors unrelated to the amount of oil or gas in the reservoir. In general, we can conclude the following:

- The amount of gas in mud may be misleading in evaluating the quality of a reservoir. A formation that is highly prospective (good saturation, porosity, and thickness) will yield relatively small

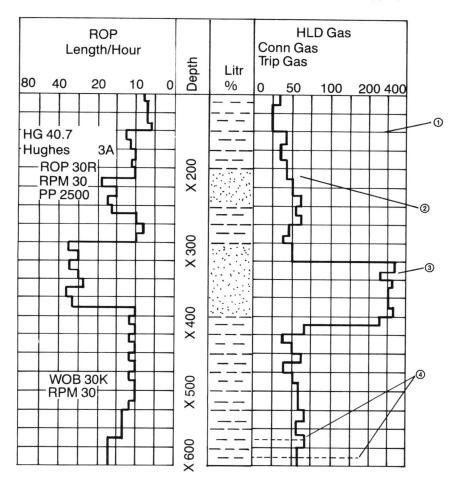

Fig. 4–5 Gas curve presentation (adapted from Anadrill's *Delta Manual*, 7-18). The ROP curve can be used with other curves as a correlation curve. Note how the curves indicate (1) trip gas, (2) drilling background, (3) gas show, and (4) connection gas.

amounts of gas in mud if the gas-oil ratio (GOR) is low. A less prospective formation where the GOR is high may produce large gas-in-mud indications. A prospective formation which is drilled very slowly with a high mud circulation ratio may yield a smaller gas-in-mud indication than a less prospective interval which is drilled rapidly.
- Complete flushing can result in no oil or gas in the mud. Vuggy

or fractured reservoirs saturated with oil that has a low GOR and residing at great depths where drilling is slow represent a worst case for successful mud logging.
- The uncertainty in evaluating flushing, retaining saturation, and losses in the return line reduce the possibility of using a numerical classification successfully in interpreting a reservoir.

COLLECTING SAMPLES

Probably one of the most important jobs of the mud logger is to collect a representative sample of drill cuttings from the shale shakers and prepare it for lithological identification and hydrocarbon show evaluation.

When the cuttings arrive at the shale shaker, they are covered in mud, unsorted by size, and generally unidentifiable. The shale shaker sifts or separates the larger cuttings from the drilling fluids and fines—microscopic or dust-sized pieces of formation. The fluid is filtered for reuse, and the cuttings are routed to the reserves pit. The mud logger collects some of the cuttings before they are routed to the reserve pit and lost. Once the samples are collected, the logger can examine them unwashed and wet, washed and wet, or washed and dried.

The unwashed samples are collected directly from the shale shakers and are untreated. They are placed in labeled sample bags and are shipped to a laboratory.

Washed samples, like unwashed, are collected from the shale shaker. However, they are washed to flush away the excess drilling mud and then are sieved to remove the coarser cavings (formation fragments that originate from the sides, not the bottom, of the borehole) before they are put in sample bags.

Washed and dried samples go through the same steps as washed samples. Before they are bagged, though, they are air dried or dried in ovens. It is from this set of samples that cuttings are taken for microscopic analysis for lithology identification and to observe oil shows.

Care must be taken to obtain representative samples, not only the last piece to arrive at the shaker. The exception is spot samples, which are taken to locate precisely the depth of a particular formation top. Once the top is measured, the balance of the samples should be collected on the shale shaker screen and a representative sample bagged. The bagged samples are sent either to the oil company or to a laboratory for analysis.

Sample Description

As we noted above, the mud logger is responsible for describing the samples. Some of the more commonly encountered rock types are described in Table 4–1.

SHOW EVALUATION

A show is the presence of hydrocarbons in a sample over and above background levels. Show evaluation is the complete analysis of the hydrocarbon-bearing formation with respect to lithology and type of hydrocarbon present. A complete show evaluation identifies the presence and type of hydrocarbon, determines the depth and thickness of the show, assesses the porosity and permeability, and assigns a show value which indicates the potential productivity of the formation.

Two types of shows are recognized: gas and oil. A gas show is hard to identify, but the mud logger may see a significant increase in gas levels. An oil show is an increase in heavier-than-methane gas levels as well as a physical indication of oil.

Identification

Four tests are used to detect hydrocarbons: odor, staining, fluorescence, and cut.

Odor does not usually apply to cuttings, but it is a useful test on cores. Although difficult to standardize, odor falls into one of four categories: poor, slight, fair, or strong.

Oil staining, like odor, is more useful when applied to cores. In general, the more desirable oils are light to colorless while viscous oils are dark. Staining is described in terms of both color and percentage of sample stained, e.g., patchy, laminated.

Liquid hydrocarbons fluoresce under ultraviolet light, and the amount, intensity, and color are the first and best indications of a show. Intensity is subdivided into none, poor, fair, or strong; color is subdivided into the values in Table 4–2.

Cut defines the leaching of oil from a sample by a solvent. Trichloroethane is the most common agent. A cut is made by taking a sample of cuttings that are fluorescing. A few drops of solvent are added. The oil leaches out of the sample and the fluorescence passes from the sample into the solvent. The rate at which the oil leaches out is classed as flash, streaming (instant, fast, slow), or crush cut. Intensity and

Table 4-1 Sample Descriptions

Argillaceous Rocks

Clay—Complex, platy alumino-silicates less than 2 microns in size. Two basic types recognized are expandable (clays that swell upon contact with water, such as montmorillonites) and nonexpandable (illites).

Claystone/Shale—Same mineral content and size as clays but indurated by compaction and dewatering. In cuttings, it is difficult to distinguish between the two. Shale must break into platelike particles.

Marl—Any clay rock (from clay to shale) with 35–65% calcareous content.

Arenaceous Rocks

Siltstone—Clay-based rock in silt-sized grains or quartz particles. Any rock or intermediate composition between clay-based and sand-based rock.

Sand/Sandstone—Pure sand grains or sand grains with a clay matrix. Grain sizes fine to very coarse and angular to rounded. Grains poorly to well sorted, cementation poor to good.

Carbonates

Limestone—Primarily calcium carbonate, recognized by fizzing strongly with 10% HCl. Some appear granular, but there are a number of classifications.

Dolomite—Similar to limestone but with a substantial part of the calcium replaced by magnesium. Less fizzing than limestone.

Evaporites

Anhydrite—Calcium sulfate or gypsum. White when pure; usually soft.

Halite (Rock Salt)—Sodium chloride. Can occur in large domes or in layers. Soft and soluble in water.

Carbonaceous Rocks

Coal—Black or dark brown, vitreous carbon. May be hard and brittle. Also occurs as peats, lignites, and other forms of organic matter.

Accessory Minerals

Pyrite—Iron sulfide. A light brassy yellow mineral associated with all sedimentary rocks. Its hardness and chemical stability may cause drilling problems if it occurs in large quantities.

Glauconite—Dark green to black iron silicate related to the mica group.

Mica—Calcium, magnesium, and iron silicate, platy in appearance.

Table 4-2 API Values of Fluorescence

API Gravity	Fluorescent color
Below 15	Brown to none
15-25	Orange to none
25-35	Yellow to cream
35-45	White
Over 45	Blue white/violet

color are also recorded. Finally, the residue that is left around the side of the spot dish is noted in white light. A dark brown, nonfluorescent ring implies bitumen, or "dead oil," rather than any producible hydrocarbons.

The type of show, whether oil or gas, needs to be determined. A gas show can be identified by an increase in total gas or by a gas chromatograph analysis showing gases heavier than methane. An oil show can be identified by one or all of these: (1) visible signs of oil on the surface of the mud, (2) fluorescence, or (3) increase of heavy hydrocarbon gases.

Depth and thickness help establish a formation's productivity on the simple premise that the thicker the show, the larger the volume of oil that can be extracted from the reservoir. In addition to these methods, porosity, permeability, and hydrocarbon ratio assist in evaluation.

Porosity

Porosity can be determined on the basis of ROP measurements. The faster the rate of penetration, the more porous the rock. This evaluation can help determine the relative porosities over the extent of the show. The first indication that a porous rock has been drilled is a drilling break. Mud loggers can also look at the samples under a microscope to make a visible estimate of porosity.

Permeability

Permeability measurements require special equipment that cannot be taken to the field, so samples must be used in the lab. The ease with which the oil is leached from the sample is an indication of its permeability. A flash cut means the oil was leached rapidly and implies good permeability. Streaming cut denotes moderate permeability; crush cut signifies poor permeability.

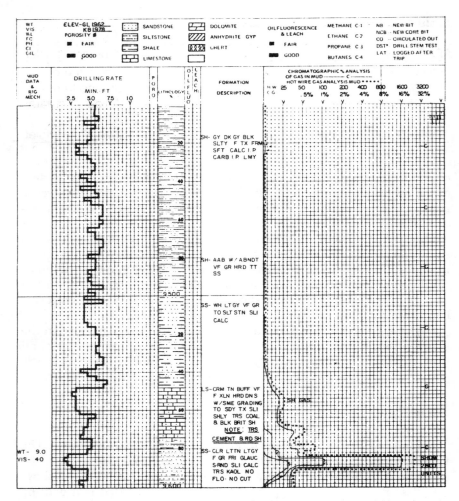

Fig. 4–6 Mud log. ROP is on the left. Both tracks show a drilling break (left) and a gas show (right) near the bottom, which indicates gas.

Hydrocarbon Ratio Analysis

Hydrocarbon ratio analysis relates the quantities of methane, ethane, propane, butane, and pentane to reservoir fluids (gas, oil, water). If the hydrocarbon ratios are plotted for each sample through a show, the gas-oil and oil-water boundaries may be established.

Application

Complete show evaluation can help us (1) identify the presence of hydrocarbons and (2) make recommendations for coring and testing programs. In coring, the standard procedure in oil companies is to circulate a drilling break up and analyze the mud and rock for signs of gas or oil. On the basis of this analysis, a core may be cut—sometimes cut continually until the oil-water contact is passed. After this, drilling may continue. In addition to coring, show evaluations correlated with offset wells and wireline logs can assist in reservoir interpretations.

The mud log (Fig. 4–6) continues to be one of the first indicators of producing formations. Once a formation is identified, further logs may be run to confirm the data—the subject of Chapter 5.

# 5

RESISTIVITY MEASUREMENT

As we saw in Chapter 3, different formations have different resistivities. More important, formation resistivity varies with porosity, water salinity, and hydrocarbon content. Although we cannot directly measure the amount of hydrocarbon in a formation, we can infer or estimate the volume of oil or gas with the aid of resistivity measurements.

Three types of logging tools are used to measure formation resistivity: induction logs, focused resistivity logs, and unfocused resistivity logs. (Note that the word "log" is used interchangeably for both the tool and the curve.) These tools can be further divided into those that measure a very small volume of the formation—microresistivity logs—and those that measure a relatively large volume of the formation.

A primitive form of the unfocused resistivity log was the first log run on an electric wireline. (Electric wireline is a wire rope with insulated electrical wires or conductors beneath the strands of cable.) This device was invented and developed by two French brothers, Conrad and Marcel Schlumberger.

As the logging industry grew, new resistivity tools were introduced that gave more accurate readings, were easier to interpret, or worked in different environments than the original electric log. Today the most common resistivity tools used are the dual induction-focused log, the dual laterolog-microfocused log, the microlaterolog, and the microlog.

Almost from the beginning, engineers realized that more than one resistivity measurement was needed because of the effects of invasion. In Fig. 5–1, the water saturation for the virgin formation is shown on the left side. At the bottom of the interval, the zone is 100% wet ($S_w = 100\%$). The resistivity of the flushed zone, R_{xo}, is 5 ohms, while the resistivity of the uninvaded zone, R_o, is 2.5 ohms. (Remember from Chapter 3 that $R_o = R_t$ when $S_w = 100\%$.)

The flushed-zone resistivity (R_{xo}) next to the borehole is higher than R_o because the mud filtrate in the flushed zone usually has a higher resistivity than the formation water resistivity (R_w) in the uninvaded portion of the formation. Occasionally, though, engineers run into a formation where the mud filtrate resistivity is less than the formation water resistivity. This may occur when a well is drilled with salt muds

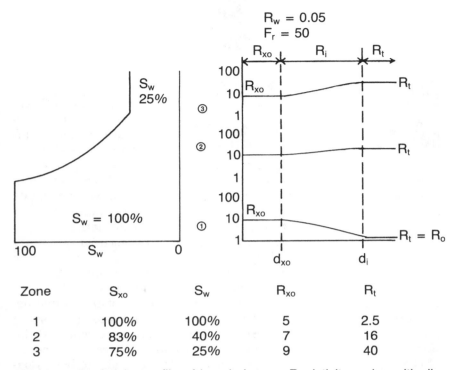

Fig. 5–1 Resistivity profile of invaded zone. Resistivity varies with distance from the borehole and the water saturation.

or at very shallow depths where the formation water is more likely to be fresh (less salty, higher resistivity).

At the top of the interval, $S_w = 25\%$. The flushed-zone resistivity (R_{xo}) shows about 9 ohms, and the true resistivity of the formation (R_t) is about 40 ohms. The R_{xo} value is higher in the top of the zone than in the bottom because some of the oil was left behind when the mud filtrate flushed the zone, and oil and gas have high resistivities. In other words, S_{xo} (the water saturation in the flushed zone) is less than 100% but greater than S_w if the zone contains oil or gas. R_t is higher than R_o because of the hydrocarbon present. The oil and gas fill some of the pore space, so there is less room for formation water. Since the resistivity of the formation depends on the amount of the formation water present (other things, such as salinity, being equal), the resistivity must increase as the volume of water decreases.

As you can see, the effects of invasion cause the resistivity to vary close to the borehole. Sometimes it is high; at other times it is low, depending on the resistivity of the mud filtrate, the formation water, the water saturation, and the porosity. The farther the measurement is taken horizontally from the borehole—into the formation—the more nearly it will match the true resistivity of the formation.

One of the problems that we encounter when using resistivity tools is that few of these tools read deeply enough to measure the formation's true resistivity (R_t) (see Table 5–1). However, because we know the approximate resistivity profile in the invaded formation and because we can measure the porosity (with other logging tools) and estimate or calculate formation water resistivity (R_w), we can construct correction curves or charts that will give a good approximation of R_t. With a good R_t we can calculate S_w, the water saturation. If S_w is low, the zone is productive; if S_w is high, the zone is wet.

INDUCTION TOOLS

The induction tool was developed to provide a means of logging wells drilled with oil-based (nonconductive) muds. All of the original electric logs used the mud column to conduct the current into the formation, so these logs could not be used in nonconductive muds or in air-drilled holes. This was a serious limitation in some areas because the water-based, conductive muds often damaged the formation by causing water-sensitive clays to swell. Formation permeability was reduced and resulted in drilling problems.

Although the induction tool was developed to meet the need for a resistivity tool that could operate in a nonconductive mud, engineers soon recognized that the tool worked better than the original electric log in freshwater muds. The induction curve was easier to read, and it read closer to true resistivity in formations where the resistivity was not over 200 ohms and R_{mf} was greater than R_w.

The induction tool works by using the interaction of magnetism and electricity. When a current is sent through a conductor, a magnetic field is created. If the current alternates, the magnetic field will also alternate by reversing poles at the same rate at which the current is alternating. We also know that if a conductor is moved through a magnetic field, a voltage is induced in the conductor. Voltage can also be induced in a stationary conductor by alternating the magnetic field. The induction tool uses these principles in its operation.

We can see this principle in action in Fig. 5–2. A high-frequency current, called the transmitter current, flows through a coil mounted in the logging sonde or tool. This current sets up a high-frequency magnetic field around the tool, extending into the formation. The con-

Table 5–1 Resistivity Tools: Uses and Limitations

Tool	Uses	Measures	VR*	DI*	Limitations
Electric	Freshwater mud, thick beds	R_t, R_i	16 in. 20 ft	16 in. 20 ft	Thin beds, difficult to interpret, obsolete
Induction	Freshwater muds, air/oil muds	R_t, R_i	5 ft	5 ft 20 ft	$R_t > 100$, $R_m < R_w$, salt muds
Dual Induction	Freshwater muds	R_t, R_i	18 in. 5 ft	30 in. 20 ft	$R_t > 100$, $R_m < R_w$, salt muds
Laterolog	Salt mud	R_t	12–32 in.	80 in.	
Dual Laterolog	Salt mud	R_t, R_{xo}	24 in.	>80 in.	
Microlog	Freshwater mud	Indicates permeability, cal	2 in.	<4 in.	Por > 15, hmc < 1/2 in.
Microlaterolog	Salt mud Freshwater mud	R_{xo}, cal	2 in.	<4 in.	$h_{mc} < 3/8$ in., inv > 4 in.
Microspherically Focused Log	Salt mud Freshwater mud	R_{xo}, cal	2 in.	<2.5 in.	

*VR = vertical resolution; DI = depth of investigation

Note: All of the major logging companies offer essentially the same measurements, although by different names. The measurements that these tools make may not be identical nor of the same accuracy, depending on the company.

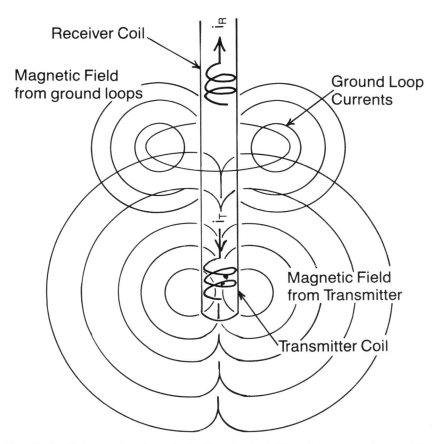

Fig. 5–2 Schematic of an induction tool. A high-frequency transmitter current induces ground currents, which in turn generate a signal in the receiver coil.

stantly changing magnetic field causes currents to flow through the formation concentric with the axis of the induction tool. The currents, called ground loops, are proportional to the conductivity of the formation; they alternate at the same frequency as the magnetic field and the transmitter current flowing through the induction coil. The ground loop currents set up a magnetic field of their own. This secondary magnetic field causes a current to flow in the receiver coil located in the induction sonde. The amount of current flowing in the receiver coil is proportional to the ground loop currents and therefore to the conductivity of the formation. The signal in the receiver coil is detected, processed, and recorded on the log as either a conductivity measurement (C) or a resistivity measurement (C = 1/R).

The tool illustrated in Fig. 5–2 is a simple two-coil device. In practice, bucking coils are used to help focus the effects of the main transmitter and receiver coils and to remove unwanted signals from the borehole. One popular induction tool used today has six different coils. The depth of investigation (the depth from which most of the measurement is obtained) for a typical deep induction tool is about 10 ft. The vertical resolution—that is, the thinnest bed that the tool will detect—is 40 in. Both the depth of investigation and the vertical resolution are affected by the spacing between the main transmitter and the receiver coils as well as by the placement of the focusing coils. By judicious selection of these parameters, we can design different depths of investigation into a tool. Thus, we can measure the resistivity profile through the invaded zone and correct the deep induction reading to move it closer to the R_t value that we want.

For many years the induction electric log was the most popular induction tool in high-porosity formations such as in California, along the Gulf Coast, and in other high-porosity, moderate-resistivity formations. A single induction curve with a vertical resolution of about 3 ft and a depth of investigation of about 10 ft was combined with either a short normal curve or a shallow laterolog curve. (These two curves will be described in the next sections.) Since invasion is seldom deep when porosity is high, these two curves, corrected for borehole and bed boundary effects, could be used to determine R_t.

The dual induction laterolog was developed for those areas that had lower porosities and deeper invasion than California-Gulf Coast-type formations. The tool has two induction curves (IL_d and IL_m) with a vertical resolution of about 40 in. However, one induction curve, the IL_d, reads very deeply into the formation, while the medium induction curve, IL_m, reads only about half as deep. A shallow-reading laterolog combined with the two induction curves gives a good description of the resistivity profile. Fig. 5–3 is a typical "tornado chart" (so called because of its distinctive shape) used to correct the dual induction-spherically focused log to obtain R_t.

Induction tools were first used in boreholes with nonconductive fluids (oil-based muds, air, and/or gas), but today they are used mostly in conductive water-based drilling muds. Although the effects of the borehole have been minimized by tool design, all induction readings are affected to a certain extent by the mud-filled borehole. The effect of the borehole is best minimized by using induction tools only when formation resistivities are less than 100 ohmmeters and when the mud filtrate resistivity, R_{mf}, is greater than the formation water resistivity, R_w. Charts correcting for the borehole signal are available from logging companies. Also see Fig. 5–4 for recommendations on which type of tool to use.

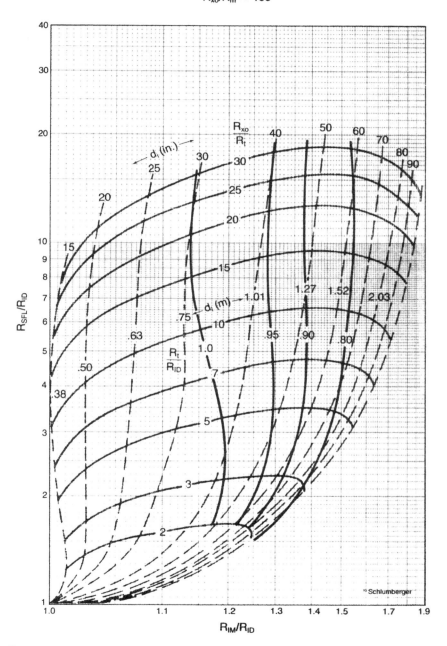

Fig. 5–3 Tornado chart (courtesy Schlumberger). Curves are used to correct deep induction resistivity to true resistivity.

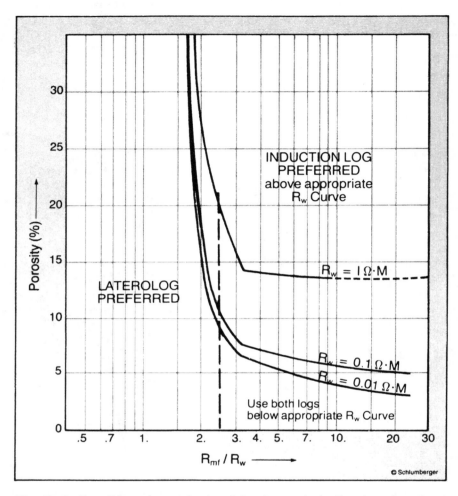

Fig. 5–4 Conditions for preferring laterolog or induction log. (courtesy Schlumberger)

FOCUSED ELECTRIC LOGS

In highly resistive formations or in very low-resistivity muds, neither the induction log nor the electric log worked very well. The mud column tended to short-circuit the current of the electric log and in these cases contributed a large portion of the total signal of the induction logs. A better measuring device than either the induction log or the electric

Resistivity Measurement 71

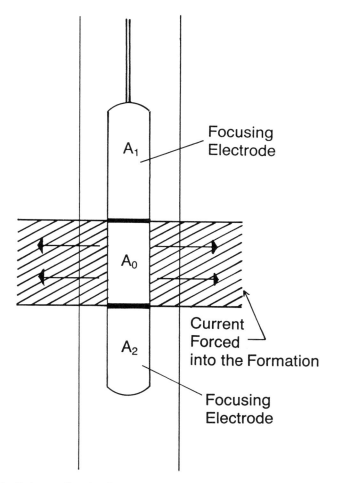

Fig. 5–5 Schematic of a focused electric log, or laterolog. A_1 and A_2 are regulated to stop any current from A_o that tries to flow up or down the borehole. Therefore all of the A_o current is forced into the formation.

log was needed for highly resistive formations, such as those in the Midcontinent and the Rocky Mountains. Focused electric logs were developed to fill this need. These tools are generally used when the R_t/R_m ratio is high, i.e., when the formation resistivity exceeds 100 ohms and/or the mud resistivity is less than the formation water resistivity (usually the case when salt muds are used).

With focused logs, the measuring current is forced out into the formation by guard electrodes. Fig. 5–5 illustrates the simplest of the focused logs, a three-electrode laterolog or guard log. Three electrodes are mounted on a sonde and are insulated from each other. The upper guard, or focusing electrode, is called A_1; the lower focusing electrode is A_2. The center electrode is the measuring electrode, A_o.

A constant current is emitted from A_o. The two guard electrodes are regulated so that any current that tries to pass them is focused and then forced into the formation instead. Therefore, low mud column resistivity or high formation resistivity has little effect on the measuring current, and accurate resistivity measurements are obtained.

Various laterolog tools have been developed over the years. Among the currently used tools, the dual laterolog is most common. This tool, similar to the dual induction log, has both a deep- and a shallow-measuring laterolog curve; it is often run in conjunction with a very shallow-reading laterolog tool which is mounted on a pad pressed against the borehole. This shallow-reading curve, called the microspherically focused log, measures the flushed-zone resistivity (R_{xo}). This combination of measurements can define the resistivity profile, which is created when mud filtrate invades the formation. Since the current path for these logs is through the mud column to the borehole wall, through the invaded zone, and then to the uncontaminated zone, the resistivity readings are a combination of these different zones. Mud and invaded zones affect the tool's resistivity measurement much less than unfocused tools, a feature which minimizes corrections. When corrections are needed, though, charts (such as Fig. 5–6) may be used to correct the resistivity readings for these effects to derive a better estimation of R_t from charts like Fig. 5–3. Refer back to Fig. 5–4 to note when induction devices are preferred to laterologs, and vice versa.

To determine which tool (induction or laterolog) to use, first estimate what the formation water will be (from a nearby well, for example). Second, estimate R_{mf}. Again, information from a nearby well can be used if the mud system will be the same. Third, calculate the ratio R_{mf}/R_w. Fourth, estimate the porosity of the zone. The numbers don't have to be exact, just reasonable guesses. For example:

Well A

$$R_w = 0.04 \text{ to } 0.08$$
$$R_{mf} = 0.2 \text{ to } 0.3$$
$$\phi = 15 \text{ to } 25\%$$
$$R_{mf}/R_w = 0.2/0.08 = 2.5 \text{ to } 0.3/.04 = 7.5$$

If the R_{mf}/R_w ratio is 2.5 to 7.5 and porosity is 15–25%, the induction log is preferred. If $R_{mf} = 0.02$ (salt mud system), $R_{mf}/R_w = 0.02/0.04 = 0.5$ or $0.02/0.08 = 0.25$ and the laterolog is preferred.

Resistivity Measurement 73

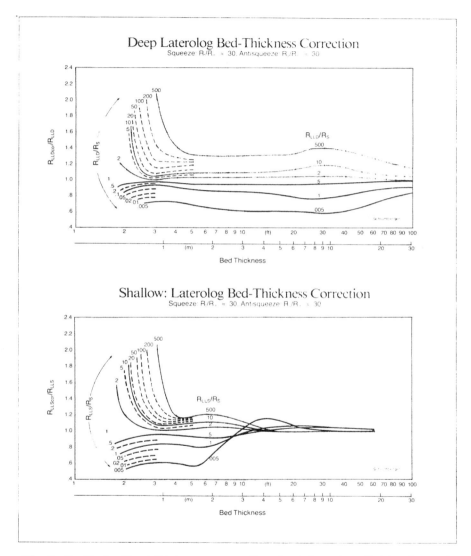

Fig. 5–6 Typical bed-thickness correction chart. (courtesy Schlumberger)

ELECTRIC LOGS

The electric log, or E-log, is of interest because so many still exist in log libraries and are used by geologists to map formations and put prospects together. Although electric logs are mechanically simple, they are very difficult to read because of the borehole effects and the many effects of the various electrode arrangements.

Fig. 5–7 illustrates a simple four-electrode system called a normal device. There are usually two normal measurements on an electric log. One has a short spacing of around 10 to 16 in. The other, called the long normal, has a spacing of 64 in.

Current from a constant source is emitted from electrode A and returns to electrode B, which is a long distance away. The current leaves A in an essentially spherical manner. The voltage at M is measured with respect to a reference electrode that is at 0 voltage. Since the current emitted by A is constant, any variation in the voltage at M will be due to changes in the resistivity of the formations (from Ohm's law, which states voltage = current × resistivity × area/length).

Another four-electrode device is called the lateral, which works on the same principle as the normal. The main difference is that three electrodes are on the sonde: A, M, and N. M and N are close together, and the voltage difference between them is measured. The advantage of the lateral over the normal is that it has a deeper radius of investigation (the depth to which it will read into the formation). This depth is about 18 ft. However, lateral curves are difficult to interpret, even for experienced log analysts.

The electric log usually consists of four measurements: spontaneous potential (SP) and three resistivity measurements with different depths of investigation—16-in. short normal, 64-in. medium normal, and 18-ft, 8-in. lateral. In some parts of the United States a 10-ft lateral was run in place of the 64-in. normal curve.

Determining true resistivity from an electric log is more an art than a science. Many rules must be followed; many corrections and judgments must be made. The thickness of the beds (formations), the mud resistivity, the shale resistivity, and the type of measuring device—normal or lateral—must all be weighed when trying to extract an accurate resistivity from these logs. Fortunately, the old E-log has been replaced by much more easily read logs, such as the induction-electric log and the laterolog.

In spite of these difficulties, do not dismiss the electric log lightly just because modern and more easily read logs are available. E-log readings often give a good indication of formation resistivity and invasion. A lot of oil and gas has been discovered with these logs.

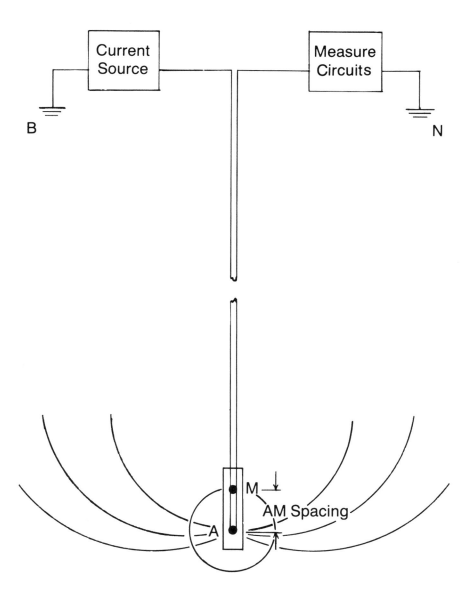

Fig. 5–7 Schematic of electric log, normal device. Current flows from electrode A to ground electrode B. Voltage is measured at electrode M with respect to another ground electrode at N.

SPONTANEOUS POTENTIAL

The spontaneous potential—SP, as it is more commonly called— is generally included on the resistivity logs. SP is not a resistivity measurement; rather, it is a naturally occurring voltage or potential (potential is an early term for voltage) caused when the conductive drilling mud contacts the formations. Since the voltage occurs naturally, it is spontaneous.

The origin and even the existence, of spontaneous potential was hotly debated in the early days of logging. Although not as crucial to log interpretations as it once was because of the abundance of other logging measurements, the SP log nevertheless can provide much useful information, especially on older logs.

The most common uses of the SP are to correlate between logs, to identify permeable formations, to measure zone thickness, to calculate R_w, and to identify shaliness. The SP log can be run only when the borehole is filled with a conductive (water-based) mud because the mud filtrate is a crucial ingredient in generating the SP voltage. The log cannot be run in oil-based mud or in air-drilled holes.

The source of the SP is the combination of the drilling mud (especially mud filtrate), the formation water, invasion, and the presence of sand and shale. The SP voltage is caused by an electrochemical action due to the differences in the salinities of the various fluids. In essence, it is a wet-cell battery similar to a car battery.

Fluids cannot flow through the shale due to the shale's very small grain size. However, due to shale's layered makeup and the negative electrical charges on these layers, sodium ions (Na^+) can flow through shale while chlorine ions (Cl^-) cannot. If we have a permeable formation such as a sandstone whose formation water contains sodium ions, and it is separated by a shale from another solution (the drilling mud) that contains a different concentration of sodium ions, current will flow because of the migration of the sodium ions.

If the drilling mud has fewer sodium ions than the formation water, current will flow from the formation water through the shale to the wellbore. This current flow is possible due to membrane potential, the voltage caused by the shale acting as a sieve which filters out everything except positively charged sodium ions.

The membrane potential accounts for about 80% of the spontaneous potential. The remainder of the SP is caused by the mud column, or mud filtrate, being in contact with the formation water. Since chlorine ions (Cl^-) are more mobile than sodium ions, more negative ions than positive will cross the junction between the two fluids. This effect is called the liquid junction potential. We add it to membrane potential to find the total SP measurement or deflection.

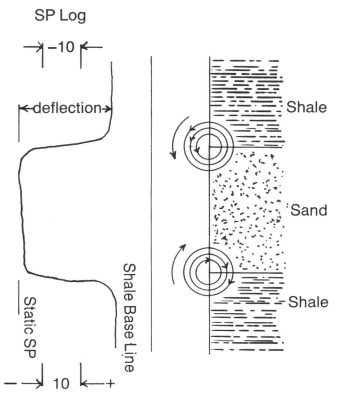

Fig. 5–8 Schematic of spontaneous potential. SP is generated by currents flowing through the mud column.

The amount of spontaneous potential in a clean (shale-free) formation is proportional to the ratio of the mud filtrate resistivity to the formation water resistivity (R_{mf}/R_w). We can use this relationship to calculate R_w.

Track 1, located to the left of the depth column, is the SP track. SP is always measured in millivolts and is often indicated as so many millivolts per division (Fig. 5–8). Note in Fig. 5–8 that the fairly constant reading opposite the shales is called the shale base line. When opposite a sandstone, the SP will normally deflect to the left (away from the depth track) opposite sands if the mud filtrate resistivity is higher than the formation water resistivity ($R_{mf} > R_w$). If a sand is thick enough, it will generally reach a constant reading that is called the static SP (SSP), the maximum SP that could be measured if no currrent were flowing in the borehole. If the sand is not thick enough or the formation resistivity is high, the SP will not reach its maximum value and corrections will have to be made.

In low-resistivity sands (for wet sands, $S_w = 100\%$) where R_t is approximately the same as the mud resistivity (R_m), the SSP value will be reached when the bed thickness (h) is about 15 times as large as the hole diameter (d). For the normal range of hole diameters, this means a bed thickness of 7–15 ft. If h is about twice as great as d, the SP curve will reach about 90% of SSP. However, in high-resistivity formations that may be hydrocarbon bearing, a much thicker formation is necessary to develop the SSP reading. If $R_t > 20 \times R_m$, then in a bed $15 \times d$ the SP will only be about 90% of SSP. We need a zone nearly 40 ft thick. For a zone only $2 \times d$ in thickness, the SP will reach less than 30% of its true value.

To have spontaneous potential, a permeable formation like a sandstone must lie next to a shale. But what happens if an impermeable, highly resistive formation like a limestone or dolomite is present? First of all, to generate spontaneous potential there must be some permeability, and carbonates usually do not have SP because they often have low permeability. Second, the high resistivity of the carbonates encourages the SP currents generated by the sands and shales to stay in the borehole fluids until they have passed the zone of high resistivity and can reach a sand or shale (see Fig. 5–9).

The presence of limestone greatly complicates interpretation of SP. Normally, it is easier to identify lithology from some other log, such as a porosity log. The R_w calculation is much less reliable because we can't always be sure we are measuring the SSP. Notice on Fig. 5–9 that the SP curve through the limestone sections tends to be a straight line connecting permeable zones. The limestone may make a formation (either sand or shale) appear thicker than it really is.

MICRORESISTIVITY TOOLS

Microresistivity tools are designed to read R_{xo}, the resistivity of the flushed zone. Since the flushed zone may be only 3 or 4 in. deep, R_{xo} tools are all very shallow reading. The electrodes are mounted on flexible pads pressed against the borehole wall, thereby eliminating most of the effects of the mud on the measurement.

The R_{xo} measurement has several uses. In certain techniques, water saturation may be obtained from resistivity ratios if S_{xo}, the flushed-zone water saturation, is known or can be estimated. An advantage of the resistivity ratio method is that water saturation can be calculated without using a porosity measurement. Another advantage is the determination of F_r from the equation: $F_r = R_{xo}/R_{mf}$.

Resistivity Measurement 79

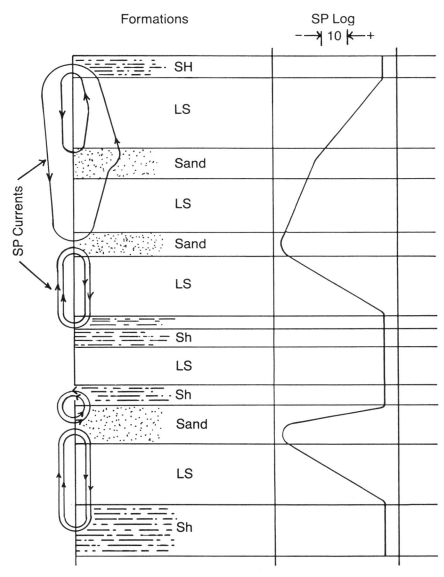

Fig. 5–9 SP curve shapes, sand/shale/limestone series.

Micrologs

The first microresistivity tool was called the microlog. On this tool, a pad carrying electrodes is filled with an insulating oil. The pad is

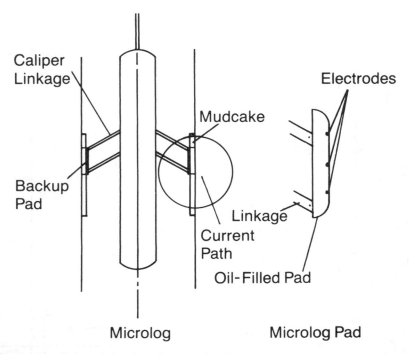

Fig. 5–10 Schematic of a microlog. Pad-mounted electrodes are pressed against the borehole wall to reduce the effects of the mud column.

pressed against the wall of the hole by the backup pad. Current flows along a path, as in Fig. 5–10.

The printed log consists of a caliper curve in track 1, which is used to measure the diameter of the borehole and to indicate the presence and thickness of mudcake, and two resistivity curves—the 1-in. × 1-in. microinverse and the 2-in. micronormal—in tracks 2 and 3. The microinverse curve has a shallower depth of investigation than the micronormal curve. The resistivity curves are closely related to the electric log's resistivity measurements but on a much smaller physical scale.

The difference in depths of investigation is used to indicate permeability. If a formation has been invaded, the mud solids will build up as mudcake on the face of the formation. The resistivity of the mudcake is usually lower than that of the formation immediately adjacent to the wellbore. Since the microinverse log makes a very shallow measurement, it will "see" primarily the mudcake. The micronormal will read more deeply into the formation and see some of the invaded zone. Therefore, the micronormal resistivity will be higher than the microinverse if the zone has been invaded. If invasion has not occurred, the two curves will read about the same.

The microlog is best used to indicate permeability and formation thickness (because of its very sharp bed boundaries) in low- to medium-resistivity formations. However, at one time, charts were used that allowed a porosity to be calculated from the microlog readings on the basis of assumed relationships between porosity, formation factor, and flushed-zone water saturation (S_{xo}). This method is still used when evaluating a set of logs that was run before the advent of porosity tools. Although micrologs are still run today, they are less important because better R_{xo} devices and much better porosity devices have been developed.

Microlaterologs

The microlaterolog (MLL) is similar in operation to its big brother, the laterolog (LL). The tool carries small concentric electrodes on a flexible pad that is pressed against the borehole wall. The outer guard electrodes force the current into the formation and prevent short-circuiting by the mudcake. For this reason, the microlaterolog is used in high-resistivity formations. A caliper log and usually a microlog are also recorded.

Microspherically Focused Logs

The microspherically focused log (MSFL) uses the same principle of operation as the spherically focused log but on a smaller scale. It is also a pad device and is often combined with other measurements such as the dual laterolog or the formation density log (to be discussed later). Like the MLL, the MSFL is recorded on a logarithmic scale.

Microresistivity tools are essential in evaluating potentially productive hydrocarbon formations. From these logs, we can determine characteristics like the following:

- depth of invasion
- flushed-zone water saturation (S_{xo})
- moved hydrocarbons ($S_{xo} - S_w$)
- corrections for deep induction and laterolog readings
- permeability
- hole diameter
- zone thickness
- porosity

In this chapter we have tied in the idea of resistivity profiles due to invasion, which we read about in Chapter 3, with methods of determining the resistivity profile. The preferred resistivity measuring tools depend on the types of formations and their resistivity. In low to medium resistivities the single or dual induction tools are used; in high resistivities the laterolog devices are best. Microresistivity tools are used to determine the resistivity of the flushed zone and the invaded zone.

We also learned about the SP curve, which is often used to compare different logs on the same well, to determine formation water resistivity (R_w), to give a sand count (net pay thickness), and to indicate permeability. The curve is normally included on all resistivity logs except microresistivity logs, and it may also be included on other logs such as the sonic, which is covered in Chapter 6.

6

POROSITY MEASUREMENTS

In Chapter 3 we discussed porosity, the space within the reservoir rocks that can contain fluids. Porosity is measured as a percentage of the bulk volume; a formation with no porosity has 0% porosity. Most reservoir rocks have a porosity range from 6 to 30%; the greater the porosity, the more fluid the rock can hold. Therefore, petroleum engineers and geologists are extremely interested in porosity.

If we could directly measure the porosity of the formations as they lie in the earth (in situ), petroleum exploration would be simple. Unfortunately, porosity is one of those variables that defies easy determination. Since we cannot measure porosity directly in the wellbore, we must deduce it from other measurements.

As accurate knowledge of porosity is important, so hundreds of thousands of research dollars have been spent on developing tools that attempt to measure it. The result of all this research is a variety of devices and techniques for measuring apparent porosity. (Apparent porosity is the porosity that a particular tool reads in a given formation.) Each of these tools determines a different apparent porosity for the same formation.

The question naturally arises, "Which one of these porosities is correct?" The answer is that all these tools can accurately determine the porosity of the reservoir rocks—under the proper conditions.

CORES

One common type of porosity measurement is core porosity. To determine core porosity, engineers cut a sample of formation, called a core. A special drilling assembly called a core barrel is lowered into the wellbore. The tool's donut-shaped bit drills a hole that leaves a solid plug of formation inside the tool's hollow center. After an appropriate distance is drilled, the tool is hauled back to the surface, retrieving the piece of cored formation inside the assembly. After the core is released from the barrel, it is packaged and sent to a laboratory

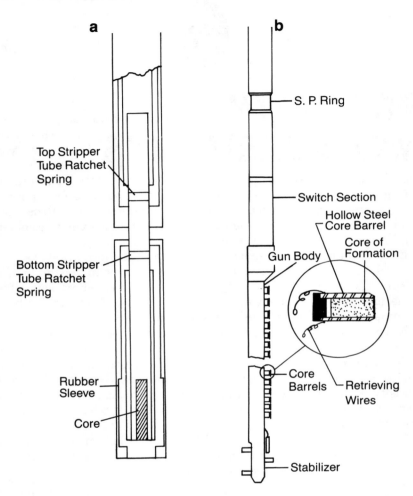

Fig. 6–1 Coring devices. (a) Core barrel run on drillpipe; (b) sidewall core gun.

where various measurements, including the porosity, are made (Fig. 6–1a).

Another method of recovering formation samples is to lower downhole a sidewall core gun, which shoots hollow steel bullets into the formation. When the bullets are retrieved, they contain samples of the formation (Fig. 6–1b). The core samples are then analyzed in a laboratory for porosity, lithology, permeability, and unusual minerals.

Core porosities may differ from the true formation porosity for several reasons: the rock's properties may be altered during the recovery process; the portion of the core that is measured may not be representative (only small plugs at intervals are actually analyzed); and the volume of rock analyzed is very small, so variations in the formation can be missed. Nevertheless, cores are the only way the driller can see the formation. With logs, he must use his imagination; but with a core, he can hold part of the underground rock in his hand.

SUBATOMIC INTERACTIONS

Today it is common to run at least two porosity devices, especially in areas with a mixed lithology of sands, limestones, dolomites, and shales. The two most popular porosity tools are the compensated density log and the compensated neutron log. Both devices use the formation's response to different types of radioactive bombardment to measure a density porosity and a neutron porosity. To understand the measurement principles on which these two tools operate, let's digress a bit for a quick lesson in nuclear physics.

The world of nuclear physics deals with matter on a scale unlike anything in common experience. The nuclear physicist works with atoms, neutrons, electrons, protons, positrons, and other subatomic units. All of these units are particles, that is, they have mass.

For the types of interactions that we'll be discussing, we need to be familiar with the following particles:

electron—has a negative charge and a very low mass
positron—just like an electron except has a positive charge
neutron—electrically neutral but has a mass almost 2,000 times that of a positron or an electron
proton—has a positive charge and essentially the same mass as a neutron
ray—nuclear particle that transports energy
alpha ray—helium atom without its electrons (2 neutrons + 2 protons)
beta ray—an electron or a positron
gamma ray—a massless particle that travels at the speed of light; also called a photon

Neutrons and protons are usually found in the nucleus of an atom (Fig. 6–2). The number of protons determines the atomic number (Z) of an element. The total number of neutrons and protons determines the atomic number (A). The number of protons and electrons is gen-

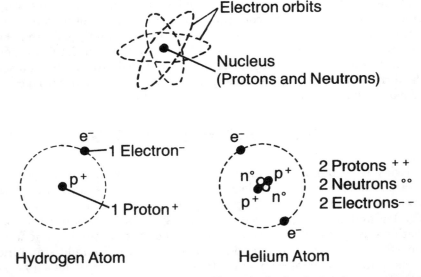

Fig. 6–2 Structure of the nucleus. The principal subatomic components are electrons, neutrons, and protons.

erally equal, which results in an electrically neutral atom. (Electrons have a low mass, so they contribute little to the weight or mass of an atom.)

If an atom is bombarded by one or more of these particles (neutrons, electrons, alpha rays, etc.), various things can happen depending on the energy of the bombarding particle, the type of particle, and the amount of energy given up to the atom. When an atom is bombarded by gamma rays, three types of interactions are possible: photoelectric effect, Compton scattering, or pair production (Fig. 6–3).

In the photoelectric effect the gamma-ray (GR) energies are less than 100 keV (thousand electron-volts). [An electron volt is a measure of energy. If a particle has an energy of 1,000 eV (1 keV), its energy is 1,000 times as great as a particle with an energy of 1 eV.] A low-energy gamma ray passes close to the nucleus of an atom, and is absorbed completely; an electron is then ejected into space. This reaction is related to the atomic number of the atom and the energy of the incident, or impinging, gamma ray. If we know the gamma ray's energy, we can make a measurement proportional to the photoelectric effect. In other words, we can approximate the atomic number, Z. Since we are dealing with a fairly limited number of atoms (primarily silicon, oxygen, calcium, hydrogen, and iron) and compounds in the reservoir rocks, we

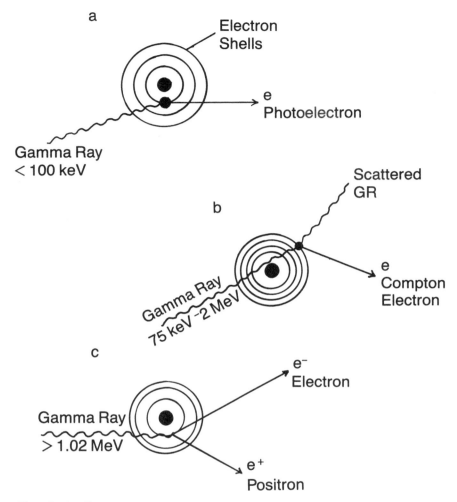

Fig. 6–3 Gamma-ray interactions. (a) Photoelectric absorption; (b) Compton scattering; (c) pair production.

can calculate the photoelectric effect for different formations. This measurement, then, is indicative of lithology and is largely unaffected by porosity.

If an incident gamma ray (GR) has an energy between 75 keV and 2 MeV (million electron-volts), the interactions between the gamma ray and the nuclei are due primarily to Compton scattering. In a Compton scattering interaction—an elastic reaction in which both energy and momentum are conserved—a gamma ray hits an electron and imparts some of its energy to that electron. (An elastic reaction is similar to the reactions between billiard balls. When the cue ball strikes another ball, the stationary ball receives some of its energy. If friction forces are disregarded, the total amount of momentum will be the same before and after the collision of the two balls. In other words, momentum of the cue ball before impact equals momentum of the cue ball after impact plus momentum of the struck ball.) The number of interactions is proportional to the number of electrons present in a unit volume. (Unit volume, you'll recall, is a cube one unit long on each side.)

Neutrons are often used as bombarding particles. They are classified according to the interactions they undergo, in much the same way as gamma rays. These interactions correspond to the following energy flevels:

- fast neutron: 100,000 to 15,000,000 electron-volts (eV)
- slow neutron: approximately 1,000 eV
- epithermal neutron: approximately 1 eV
- thermal neutron: approximately 1/40 eV

The neutrons used in logging come from a source carried in the logging tool. These sources contain types of radioactive material which naturally emit "fast" neutrons that have elastic reactions while they are in the higher energy ranges. However, with each reaction the neutron loses some of its energy. As a result, the neutron may pass through all of the stages—slow neutron, epithermal neutron, and finally thermal neutron—before it finally loses enough energy to be captured by an atom.

Capture is the other type of interaction that a neutron may undergo. When the neutron is absorbed into or captured by an atom, the atom becomes highly excited (energized) and releases this energy by emitting a gamma ray. This kind of ray is called a gamma ray of capture.

The elastic scattering reaction of the high-energy neutrons is sometimes called the colliding ball reaction. Imagine that the speeding neutron collides with a stationary atom. If the mass of the atom is much greater than the mass of the neutron, the neutron will simply bounce off and lose very little energy, like when a golf ball is dropped onto a sidewalk. The massive sidewalk hardly moves at all, but the lightweight golf ball bounces nearly as high as it was dropped; it loses very little of its energy.

On the other hand, if the neutron should collide with something that has essentially the same mass as the neutron, most of its energy will be transferred to the object it strikes. This is like hitting a cue ball into a billiard ball. If the billiard ball is struck straight on, most of the energy will be imparted to it and the cue ball will stop dead.

A neutron has very nearly the same mass as a hydrogen atom. Therefore, the amount of energy that a neutron loses is proportional to the number of hydrogen atoms present. After a few collisions, the neutron is slowed enough to be absorbed or captured by a nearby nucleus. The nucleus then emits a gamma ray of capture. By counting these capture GRs, we make a measurement that is proportional to the number of hydrogen atoms present.

DENSITY LOGS

Density is the weight of a unit volume of a substance. For example, 1 cu ft of distilled water weighs 62.4 lb and thus has a density of 62.4 lb/cu ft in the English system of measurement, while pure limestone's density is 169 lb/cu ft. In the metric measuring system 1 cc of water weighs 1 g, so water has a density of 1.0 g/cc and limestone's density is 2.71 g/cc. In logging in the United States, we use the centimeter-gram-second metric system for density measurements, i.e., g/cc. In many parts of the world, the SI metric system is used.

Unfortunately, we can't measure formation, or bulk, density directly from the borehole. However, we can measure electron density by using Compton scattering reactions, and electron density is nearly the same measurement as bulk density.

The density-measuring tool (Fig. 6–4) bombards the formation adjacent to the wellbore with gamma rays from a cesium source, and Compton scattering takes place. The gamma rays are counted by two detectors mounted on a skid pressed against the borehole wall. The two detectors allow compensation for the effects of hole roughness and mudcake while the bulk density measurement is made. (That's why the tools are called compensated density tools.)

The newest generation of density tools measures the photoelectric absorption cross section of the formations. The photoelectric effect is simply another way the formation reacts to the bombarding gamma rays. This reaction occurs at a much lower energy level than Compton scattering. By measuring the energy level of the formation reactions, engineers can separate photoelectric reactions from other reactions. The photoelectric response is then used to help identify lithology.

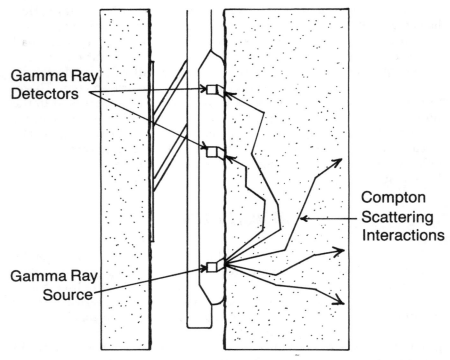

Fig. 6-4 Compensated formation density tool. Gamma rays emitted by the source undergo Compton scattering reactions with the formation. The two detectors compensate for hole rugosity.

Interpretation—Density Log

The density determined by the density tool is called ρ_b, where b stands for bulk volume. The rock structure of minerals such as sandstone or limestone is called the matrix. The density of this structure is called the matrix density, ρ_{ma}. This is the density that the tool would read if the formation had zero porosity. The density of the fluid, usually mud filtrate, in the pore space is called ρ_f.

A very important principle of mathematics is that the whole is equal to the sum of its parts. We use this axiom constantly in log interpretation; the determination of porosity from density measurements is a good illustration. The bulk volume of formation is the whole, and the parts are the matrix volume and the fluid volume contained in the pore space. Let's assume that we know the matrix density (the densities of sandstones, limestones, dolomites, and other minerals have been measured in the laboratory) and the fluid density. Then we can write an equation based on the principle that the whole equals the sum of its parts:

density log reading (ρ_b) = matrix volume (BVM) × matrix density (ρ_{ma}) + fluid volume (ϕ) × fluid density (ρ_f)

But we know that the matrix volume = $(1 - \phi)$ and the fluid volume = ϕ, so:

$$\rho_b = (1 - \phi) \times \rho_{ma} + \phi \times \rho_f$$

To apply this principle to the earth's formations in order to determine porosity, we must know or assume (1) the density of the matrix and (2) the density of the fluid in the pore spaces. We can often make these assumptions in developed fields where the lithology is accurately known or in regions that are predominantly sandstone such as the U.S. Gulf Coast or California.

We know from experience and laboratory measurements that the matrix density is 2.71 g/cc for limestone, 2.87 g/cc for dolomite, and for sandstones 2.65 g/cc (unconsolidated) or 2.68 g/cc (mature). The fluid in the pore space is water, oil, or gas. Since the density tool has a shallow depth of investigation, fluid density is generally assumed to be 1.0 g/cc, but corrections can be applied to the value if necessary.

Fig. 6–5 is a section of a Litho-Density log (mark of Schlumberger). Note the scales and curves in tracks 2 and 3. On the log, the density curve (solid) is designated RHOB (ρ_b) and is scaled from 2.0 to 3.0 g/cc. The dashed curve labeled PEF and scaled 0.0 to 10.0 is the photoelectric index curve. The dotted curve in track 3 is the density correction curve (DRHO for $\Delta\rho$). It monitors the amount of correction being added to the ρ_b curve by the compensation circuitry. If $\Delta\rho$ is greater than 0.15, use the density reading with caution because the correction is excessive and the density reading may be incorrect. The heavy dashed line at division 8 of track 3 is the tension curve, which monitors the drag when the logging tool rubs against the side of the hole.

To use the density curve, we need a chart that converts ρ_b to porosity (Fig. 6–6). To use the chart we need to know the lithology. That's where the PEF curve comes in. Since the PEF for sandstone is 1.8, limestone is 5.1, and dolomite is 3.1 (Table 6–1), we can use the PEF curve to identify the matrix minerals. The only obstacle is shale, whose PEF can range from 1.8 to 6.3 but usually is around 3, like that of dolomite.

text continued on page 98

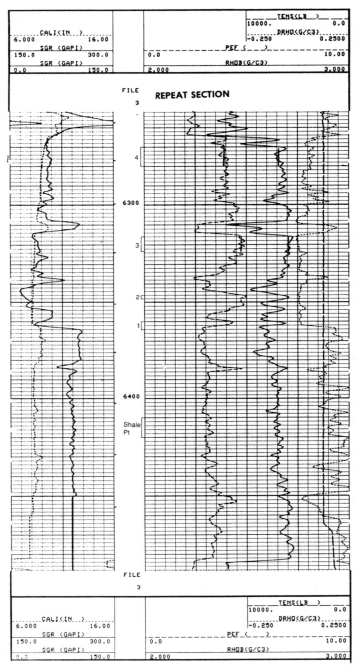

Fig. 6−5 Section of a Litho-Density log. (mark of Schlumberger)

Porosity Measurements 93

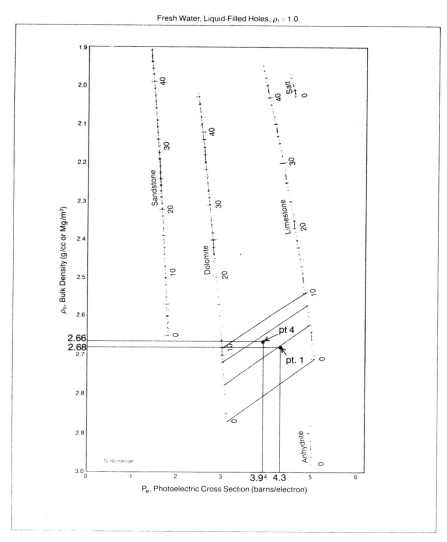

Fig. 6–6 Chart to determine porosity and lithology from Litho-Density log. (courtesy Schlumberger)

Table 6-1 Logging Tool Response in Sedimentary Materials

Name	Formula	ρ_{LOG} g/cc	ϕ_{SNP} p.u.	ϕ_{CNL} p.u.
CLAYS				
Kaolinite	$Al_4Si_4O_{10}(OH)_8$	2.41		37.
Chlorite	$(Mg,Fe,Al)_6(Si,Al)_4O_{10}(OH)_8$	2.76	37.	52.
Illite	$K_{1-1.5}Al_4(Si_{7-6.5},Al_{1-1.5})O_{20}(OH)_4$	2.52	20.	30.
Montmorillonite	$(Ca,Na)_7(Al,Mg,Fe)_4$ $(Si,Al)_8O_{20}(OH)_4(H_2O)_n$	2.12	40.	44.
EVAPORITES				
Halite	NaCl	2.04	-2.	-3.
Anhydrite	$CaSO_4$	2.98	-1.	-2.
Gypsum	$CaSO_4(H_2O)_2$	2.35	50+	60+
Trona	$Na_2CO_3NaHCO_3H_2O$	2.08	24.	35.
Tachydrite	$CaCl_2(MgCl_2)_2(H_2O)_{12}$	1.66	50+	60+
Sylvite	KCl	1.86	-2.	-3.
Carnalite	$KClMgCl_2(H_2O)_6$	1.57	41.	60+
Langbenite	$K_2SO_4(MgSO_4)_2$	2.82	-1.	-2.
Polyhalite	$K_2SO_4MgSO_4(CaSO_4)_2(H_2O)_2$	2.79	14.	25.
Kainite	$MgSO_4KCl(H_2O)_3$	2.12	40.	60+
Kieserite	$MgSO_4H_2O$	2.59	38.	43.
Epsomite	$MgSO_4(H_2O)_7$	1.71	50+	60+
Bischofite	$MgCl_2(H_2O)_6$	1.54	50+	60+
Barite	$BaSO_4$	4.09	-1.	-2.
Celesite	$SrSO_4$	3.79	-1.	-1.
SULFIDES				
Pyrite	FeS_2	4.99	-2.	-3.
Marcasite	FeS_2	4.87	-2.	-3.
Pyrrhotite	Fe_7S_8	4.53	-2.	-3.
Sphalerite	ZnS	3.85	-3.	-3.
Chalcopyrite	$CuFeS_2$	4.07	-2.	-3.
Galena	PbS	6.39	-3.	-3.
Sulfur	S	2.02	-2.	-3.

(Courtesy Schlumberger)

Porosity Measurements 95

t_c μs/ft	t_s μs/ft	P_e barn/elect	U barn/cc	ε farads/m	t_p nsec/m	GR API units	Σ c.u.
		1.83	4.44	~5.8	~8.0	80–130	14.12
		6.30	17.38	~5.8	~8.0	180–250	24.87
		3.45	8.73	~5.8	~8.0	250–300	17.58
		2.04	4.04	~5.8	~8.0	150–200	14.12
67.0	120.	4.65	9.45	5.6–6.3	7.9–8.4	—	754.2
50.		5.05	14.93	6.3	8.4	—	12.45
52.		3.99	9.37	4.1	6.8	—	18.5
65.		0.71	1.48			—	15.92
92.		3.84	6.37			—	406.02
		8.51	15.83	4.6–4.8	7.2–7.3	500+	564.57
		4.09	6.42			~220	368.99
		3.56	10.04			~290	24.19
		4.32	12.05			~200	23.70
		3.50	7.42			~245	195.14
		1.83	4.74			—	13.96
		1.15	1.97			—	21.48
100.		2.59	3.99			—	323.44
		266.82	1091.			—	6.77
		55.19	209.			—	7.90
39.2	62.1	16.97	84.68			—	90.10
		16.97	82.64			—	88.12
		20.55	93.09			—	94.18
		35.93	138.33	7.8–8.1	9.3–9.5	—	25.34
		26.72	108.75			—	102.13
		1631.37	10424.			—	13.36
122.		5.43	10.97			—	20.22

Continued.

Table 6-1 Logging Tool Response in Sedimentary Materials—cont'd

Name	Formula	ρ_{LOG} g/cc	ϕ_{SNP} p.u.	ϕ_{CNL} p.u.
COALS				
Anthracite	$CH_{.358}N_{.009}O_{.022}$	1.47	37.	38.
Bituminous	$CH_{.793}N_{.015}O_{.078}$	1.24	50+	60+
Lignite	$CH_{.849}N_{.015}O_{.211}$	1.19	47.	52.
SILICATES				
Quartz	SiO_2	2.64	−1.	−2.
β-Cristobalite	SiO_2	2.15	−2.	−3.
Opal (3.5% H_2O)	$SiO_2 (H_2O)_{.1209}$	2.13	4.	2.
Garnet	$Fe_3Al_2(SiO_4)_3$	4.31	3.	7.
Hornblende	$Ca_2NaMg_2Fe_2AlSi_8O_{22}(O,OH)_2$	3.20	4.	8.
Tourmaline	$NaMg_3Al_6B_3Si_6O_2(OH)_4$	3.02	16.	22.
Zircon	$ZrSiO_4$	4.50	−1.	−3.
CARBONATES				
Calcite	$CaCO_3$	2.71	0	−1.
Dolomite	$CaCO_3MgCO_3$	2.88	2.	1.
Ankerite	$Ca(Mg,Fe)(CO_3)_2$	2.86	0	1.
Siderite	$FeCO_3$	3.89	5.	12.
OXIDATES				
Hematite	Fe_2O_3	5.18	4.	11.
Magnetite	Fe_3O_4	5.08	3.	9.
Geothite	$FeO(OH)$	4.34	50+	60+
Limonite	$FeO(OH)(H_2O)_{2.05}$	3.59	50+	60+
Gibbsite	$Al(OH)_3$	2.49	50+	60+
PHOSPHATES				
Hydroxyapatite	$Ca_5(PO_4)_3OH$	3.17	5.	8.
Chlorapatite	$Ca_5(PO_4)_3Cl$	3.18	−1.	−1.
Fluorapatite	$Ca_5(PO_4)_3F$	3.21	−1.	−2.
Carbonapatite	$(Ca_5(PO_4)_3)_2CO_3H_2O$	3.13	5.	8.
FELDSPARS—Alkali				
Orthoclase	$KAlSi_3O_8$	2.52	−2.	−3.
Anorthoclase	$KAlSi_3O_8$	2.59	−2.	−2.
Microcline	$KAlSi_3O_8$	2.53	−2.	−3.

t_c μs/ft	t_s μs/ft	P_e barn/elect	U barn/cc	ϵ farads/m	t_p nsec/m	GR API units	Σ c.u.
105.		0.16	0.23			—	8.65
120.		0.17	0.21			—	14.30
160.		0.20	0.24			—	12.79
56.0	88.0	1.81	4.79	4.65	7.2	—	4.26
		1.81	3.89			—	3.52
58.		1.75	3.72			—	5.03
		11.09	47.80			—	44.91
43.8	81.5	5.99	19.17			—	18.12
		2.14	6.46			—	7449.82
		69.10	311.			—	6.92
49.0	88.4	5.08	13.77	7.5	9.1	—	7.08
44.0	72.	3.14	9.00	6.8	8.7	—	4.70
		9.32	26.65			—	22.18
47.		14.69	57.14	6.8–7.5	8.8–9.1	—	52.31
42.9	79.3	21.48	111.27			—	101.37
73.		22.24	112.98			—	103.08
		19.02	82.55			—	85.37
56.9	102.6	13.00	46.67	9.9–10.9	10.5–11.0	—	71.12
		1.10				—	23.11
42.		5.81	18.4			—	9.60
42.		6.06	19.27			—	130.21
42.		5.82	18.68			—	8.48
		5.58	17.47			—	9.09
69.		2.86	7.21	4.4–6.0	7.0–8.2	~220	15.51
		2.86	7.41	4.4–6.0	7.0–8.2	~220	15.91
		2.86	7.24	4.4–6.0	7.0–8.2	~220	15.58

Continued.

Table 6-1 Logging Tool Response in Sedimentary Materials

Name	Formula	ρ_{LOG} g/cc	ϕ_{SNP} p.u.	ϕ_{CNL} p.u.
FELDSPARS—Plagioclase				
Albite	$NaAlSi_3O_8$	2.59	−1.	−2.
Anorthite	$CaAl_2Si_2O_8$	2.74	−1.	−2.
MICAS				
Muscovite	$KAl_2(Si_3AlO_{10})(OH)_2$	2.82	12.	20.
Glauconite	$K_2(Mg,Fe)_2Al_6(Si_4O_{10})_3(OH)_2$	~2.54	~23.	~38.
Biotite	$K(Mg,Fe)_3(AlSi_3O_{10})(OH)_2$	~2.99	~11.	~21.
Phlogopite	$KMg_3(AlSi_3O_{10})(OH)_2$			

Shale has a slight effect on the density porosity readings, so its influence must be removed. To remove the shale effect:

1. Find a uniform shale section
2. Read the apparent density porosity in the shale
3. Apply the shale correction equation:

$$\phi_{Dcor} = \phi_D - (V_{sh} \times \phi_{Dsh})$$

Where:

ϕ_{Dcor} = corrected density porosity
ϕ_D = density porosity from the log
V_{sh} = shale volume
ϕ_{Dsh} = apparent density porosity in the shale

In practice, the effect of shale on the density log is often ignored unless V_{sh} is high (> 30%).

Now it's time for some practice in converting density readings to porosities. Set up a table so you can record the values from the log in Fig. 6–5. The table should look like this:

Point No.	Pef	ρ_b	$\Delta\rho$	Porosity	Lithology
1	4.3	2.69	+0.01		
2					
3					
4					

Now read the values (except porosity—you'll find that from the chart) off the log at these points. You're already started with point 1.

t_c μs/ft	t_s μs/ft	P_e barn/ elect	U barn/cc	ε farads/m	t_p nsec/m	GR API units	Σ c.u.
49.	85.	1.68	4.35	4.4–6.0	7.0–8.2	—	7.47
45.		3.13	8.58	4.4–6.0	7.0–8.2	—	7.24
49.	149.	2.40	6.74	6.2–7.9	8.3–9.4	~270	16.85
		6.37	16.24				24.79
50.8	224.	6.27	18.75	4.8–6.0	7.2–8.1	~275	29.83
50.	207.						33.3

If you have trouble reading the log, review Chapter 2. If you just want to check your answers, see the end of this chapter for our picks, complete with porosity and lithology estimate.

Once you've made your log picks, turn to Fig. 6–6. Enter point 1 at the bottom of the chart at PEF = 4.3. Draw a vertical line. Next enter $ρ_b$ = 2.69 at the left edge and draw a horizontal line until it intersects the PEF line. This line plots between the dolomite line and the limestone line. To determine the porosity, connect points of equal porosity on the two matrix lines (10 to 10, 5 to 5, etc.). Point 1 falls on the 5% line and is closer to the limestone line than to the dolomite line, so enter 5% and dolomitic limestone in the table.

Point No.	Pef	$ρ_b$	Δρ	Porosity	Lithology
1	4.3	2.69	+0.01	5%	Dolomitic limestone
2					
3					
4					

Try your hand at the other three points; then check your answers.

Although adding the PEF curve to the density tool helps determine both porosity and lithology under favorable conditions, sometimes the PEF curve isn't available (slim tools), doesn't work (heavy mud weights), or wasn't run (older density tools). In those cases we either assume the lithology or we must have additional information that can only be provided by another porosity measurement. One of these tools is the compensated neutron log.

COMPENSATED NEUTRON LOG

The original neutron tool was an early development. By bombarding the formation with neutrons from a chemical source in the logging tool, engineers could measure the response of the formation as a function of the number of hydrogen atoms present. Because most of the hydrogen present is in the water (H_2O) and oil (C_2H_{2n+2}) and because one or both of these fluids are present in the pores of the rocks, we can determine the porosity simply by counting the hydrogen atoms.

This is what the single-detector neutron tool does. However, it indicates rather than measures porosity. The response of the tool is nonlinear; it has high resolution in low porosities but very little resolution in high porosities. For this reason, it is used mainly in hard-rock areas and as a correlation tool in casing. (Casing prevents most tools from making a valid measurement, but both the natural gamma-ray and the neutron tool will read through the casing. Often the two measurements are used to correlate the open-hole measurements to some feature of the casing, such as the depth of the casing collars. The casing collars, which are easily detected, may then be used as a reference for positioning perforating guns, plugs, packers, etc.)

In the early '70s a major development was made in neutron tool instrumentation, resulting in the compensated neutron (CN) device. The CN uses two detectors to compensate for hole rugosity or roughness. In addition, it measures the ratio of the detector responses and converts this ratio to a linear porosity reading instead of the nonlinear response of the single-detector neutron tool.

Interpretation—Compensated Neutron Log

The compensated neutron log is generally recorded on a limestone matrix. Thus the porosity readings shown on the log are correct if:

- The formation is clean (no shale)
- The porosity is filled with liquid (not with gas)
- The formation is limestone

Remember that the neutron tool is responding principally to the number of hydrogen atoms present. Also remember that shale has a large number of hydrogen atoms because water molecules are bound to the clay, but its effective porosity is essentially zero due to the very fine grain size. The presence of shale in a formation increases the total porosity, but the effective porosity remains the same. Since we're interested in effective porosity and not total porosity, we must subtract the shale value from the total porosity reading. We do this by applying the following equation:

$$\phi_{Ncor} = \phi_N - (V_{sh} \times \phi_{Nsh})$$

Where:

ϕ_{Ncor} = corrected neutron porosity
ϕ_n = apparent neutron porosity read from the log
V_{sh} = shale volume
ϕ_{Nsh} = apparent neutron porosity at the shale point

The neutron porosity should be corrected for shale whenever V_{sh} exceeds 5%.

The presence of gas in the formation also has a pronounced effect on the neutron log readings. Since gas has many fewer hydrogen atoms per unit volume than either water or oil, the apparent porosity in a gas zone is much lower than it should be. To correct for the gas effect, we need to know density and/or sonic porosity.

If the formation is not limestone but sandstone or dolomite, we must correct the apparent limestone porosity to the proper matrix using Fig. 6–7. However, we can use this chart only if we know the matrix. If we don't know the matrix, we need another porosity measurement to determine porosity and lithology.

Another type of neutron porosity tool is the SNP or sidewall neutron porosity device, so called because the detector and the source are mounted on a skid similar to the density skid and are pressed against the side of the borehole. One of the advantages of the SNP is that it can be run in empty or air-filled holes. The CN log does not respond properly in these conditions, but the SNP does. Before the development of the CNL, the SNP was often used in mud-filled holes. However, today the SNP is generally used only in air-drilled holes.

SONIC LOG

The sonic or acoustic tool uses sound waves to measure porosity. Let's take a quick look at sound waves and how they travel.

Sound is energy that travels in the form of a wave and has a frequency between 20 and 20,000 cycles per second (cps, or Hertz). A sound wave (also called an acoustic wave) can travel in several different forms. The most common form is a compressional wave, the kind of wave that vibrates our eardrums so we can hear. Compressional waves are also called P-waves (primary waves) because they are the first waves to arrive.

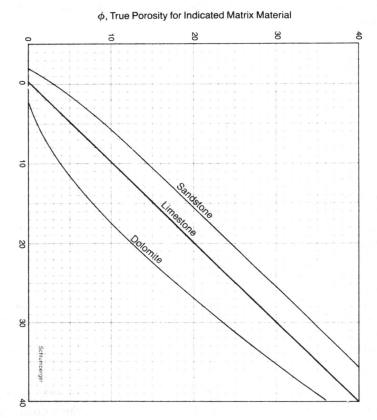

Fig. 6–7 Neutron porosity equivalence curves (courtesy Schlumberger). Usually the neutron log is recorded in limestone porosity units (pu's). Use the chart to convert to other rock types.

A compressional wave travels by compressing the material in which it travels. The material "moves" along the axis of the wave. An example of a compressional or P-wave is a Slinky™ spring toy that you hold outstretched vertically. If you lift a couple of coils and then drop them, a compressional wave will travel down the spring. When the wave reaches the end of the coil, it will travel back up. This phenomenon is called reflection.

Another characteristic of sound waves is that they change speed when the material in which they are traveling changes. This process is called refraction.

A second type of sound wave is the shear wave or S-wave. This wave is slower than the P-wave and cannot be transmitted through a fluid.

To visualize an S-wave, think of a rope with one end tied to a tree. If you pull the free end of the rope almost tight and then snap it, a shear wave will roll down the rope. The rope does not move horizontally; it moves vertically, or at right angles to the axis of the wave. This motion is characteristic of a shear wave. If it were not in a solid medium, it would not be able to transmit its energy.

Several other types of waves may be present in a full-wave recording of sound passing through the formation near the borehole. These waves are of little practical importance at present, although researchers are investigating them.

The sonic tool takes advantage of the fact that a sound wave travels at different speeds through different materials and, more important, that the sound wave travels at different speeds through mixtures of materials. If we know the speed of sound for each of the materials, we can calculate the amount of each material as long as there are only two materials. If there are more than two substances, we need additional information. In other words, if we know that a certain formation is a limestone and that any pore spaces it may have are filled with water, we can determine the porosity by measuring the time a compressional sound wave takes to travel through 1 ft of the formation.

Fig. 6–8 is a schematic of a sonic tool that has one transmitter and two receivers. The transmitter is 3 ft from the first receiver and 5 ft from the second receiver, and it emits a strong sound pulse that travels spherically outward in all directions. The mud column and the tool have slower travel times (sonic velocities) than the formations.

The first sound energy to arrive at the two receivers is the compressional wave (P-wave), which travels through the formation near the borehole. The difference in the times at which the signal reaches the two receivers is divided by the spacing of the receivers. This time, recorded in microseconds per foot, is also called sonic interval transit time (t) for the difference in arrival times between the two receivers.

In practice, the tools that measure sonic interval transit times or t are much more complicated than the tool illustrated here. They have multiple transmitters and receivers to compensate for sonde tilt, washed-out hole, and alteration of the rock properties near the wellbore due to drilling processes.

Interpretation—Sonic Log

If we know the interval transit time and the type of formation and can assume the porosity is uniformly distributed (intergranular as opposed to vugular or fracture porosity), we can determine the porosity

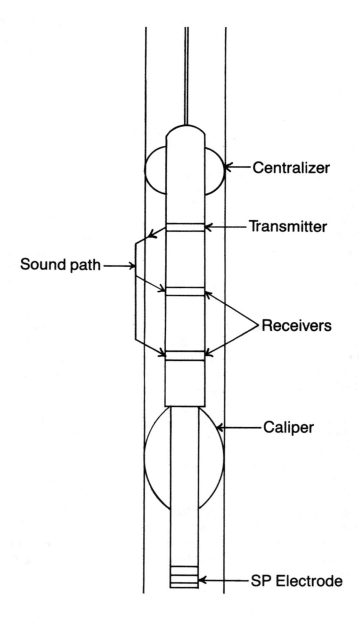

Fig. 6–8 Simplified schematic of a borehole-compensated sonic tool. The upper and lower transmitters eliminate the effects of washouts on the measurement.

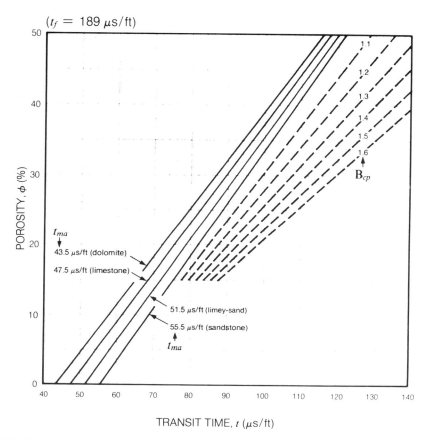

Fig. 6–9 Determining porosity from sonic interval transit time (courtesy Schlumberger). Sonic interval transit times (t) may be converted to porosity if the lithology is known.

(Fig. 6–9). As with any single porosity measurement, we must know or assume the lithology to make an estimate.

Shale has a strong effect on the sonic log. In shaly formations ($V_{sh} > 5\%$) the sonic porosity must be corrected for the presence of shale with the equation:

$$\phi_{Scor} = \phi_s - V_{sh} \times \phi_{Ssh}$$

Where:

ϕ_{scor} = corrected sonic porosity
ϕ_s = sonic porosity determined from Fig. 6–9

V_{sh} = shale volume
ϕ_{Ssh} = apparent porosity of shale point

Gas also has a strong effect on the apparent sonic porosity: it raises the apparent porosity. If gas-bearing formations are anticipated, engineers should run at least one other porosity device, preferably a compensated neutron log.

Unconsolidated sandstones such as those in California or on the Gulf Coast have longer travel times than they should for their porosity. To correct the travel times for unconsolidation, the t reading at the shale point is used to determine the compaction correction (B_{cp}):

$$B_{cp} = t_{sh}/100$$

For example, if t_{sh} = 120 μs/ft, then B_{cp} is 120/100 = 1.2.

Instead of the sandstone line (55 μs/ft) in Fig. 6–9, the correct B_{cp} line is used for entering with t from the log when the formation is unconsolidated.

Finally, the sonic wave does not "see" the porosity in vugs and fractures (secondary porosity) as well as it "sees" intergranular (primary) porosity. This lowers the apparent porosity in vugular and/or fractured formations. By comparing the sonic primary porosity to the total porosity from other logs, we can estimate the amount of fracture or vugular (secondary) porosity.

MULTIPLE POROSITY LOGS

We have already seen how to derive porosity from the individual porosity logs. With a single device we must assume we know the mineralogy of the formation, but unfortunately this is seldom the case. Both gas and shale make the interpretation of single porosity measurements less reliable. Fortunately, a powerful technique exists that helps us overcome many of the limitations of the single porosity measurement: the porosity cross-plot technique.

With the cross-plot technique and two porosity measurements, we can determine porosity that is independent of lithology, i.e., we don't have to assume the lithology. Since three different porosity tools are available (density, neutron, and sonic), we can combine cross-plots. These plots, called mineral identification plots, allow a very accurate estimate of rock type. The most common combination is made with the compensated neutron-density log.

When the density and the neutron logs are run together, they are usually recorded on a limestone matrix as though all the formations were limestones. (It is very important to check the matrix type used to record the logs—the matrix varies from region to region.) In a shale-free, wet limestone, the two logs will read the same porosity. If the zone is a wet sandstone, the neutron porosity will read too low and the density porosity will read too high. There will be a positive separation between the two curves, which is always of interest. The separation could be due to a different matrix or to the presence of gas. Usually if the separation is more than six porosity units (pu's), the reason is gas. To be certain, a third porosity measurement is necessary (or the lithology must be assumed). Shale and dolomite also cause the density and neutron porosity curves to separate, but in the opposite direction from gas or sandstone. In a shale or dolomite, the neutron porosity will be higher than the density porosity. A third porosity device, such as the sonic, is usually necessary to determine whether the separation is due to shale or dolomite.

With three porosity logs, lithology as well as porosity can be accurately determined. Once lithology is known, the density and the neutron values can be corrected for matrix and entered in Fig. 6–10.

Density and neutron tools are often "stacked" (connected so they can be run simultaneously) when logging wells with unknown or mixed lithology or in gas wells. Fig. 6–11 is the neutron-density log over the same interval of formation that we examined with the Litho-Density tool in Fig. 6–6. By entering the neutron (ϕ_N) and density (ϕ_D) porosities in Fig. 6–12, we can estimate the lithology and plot the porosity.

Let's try the same points as before. This time, though, we must correct for shale on both the density and neutron readings. Remember:

$$\phi_{Ncor} = \phi_N - (V_{sh} \times \phi_{Nsh})$$

$$\phi_{Dcor} = \phi_D - (V_{sh} \times \phi_{Dsh})$$

Note that the density porosity (ϕ_D) is the solid curve and the neutron porosity (ϕ_N) is the dashed curve. Both curves are recorded on a limestone matrix and are scaled 30% to −10%. First let's set up a table.

Point No.	Vsh, %	ϕ_D, %	ϕ_N, %	ϕ_{Dcor}, %	ϕ_{Ncor}, %	ϕ_{XP}, %	Lithology
Shale	100	3.5	25	—	—	—	Shale
1	5	1	2	1	1	1	Limestone
2	0						
3	25						
4	50						

108   Well Logging for the Nontechnical Person

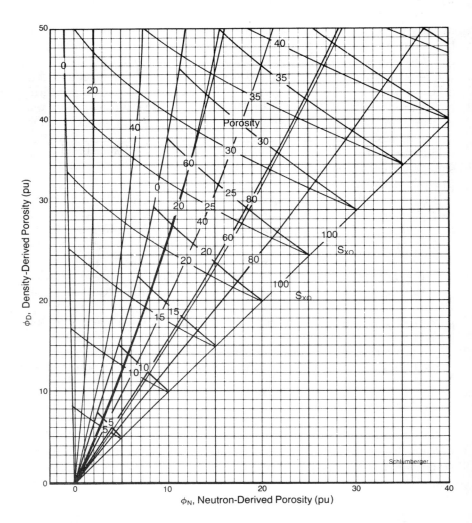

Fig. 6–10 Gas-bearing formation porosity from density and neutron logs. (courtesy Schlumberger)

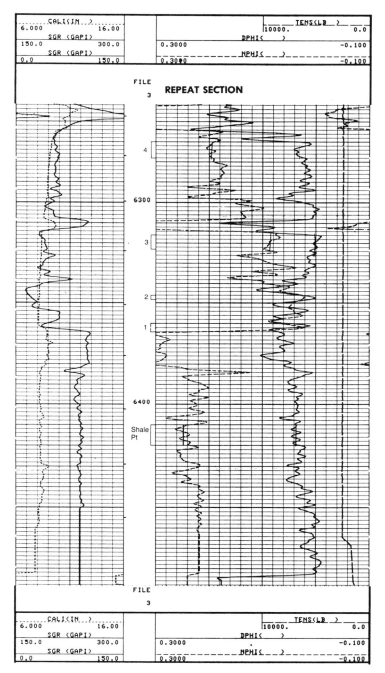

Fig. 6–11 Sample compensated neutron-density log.

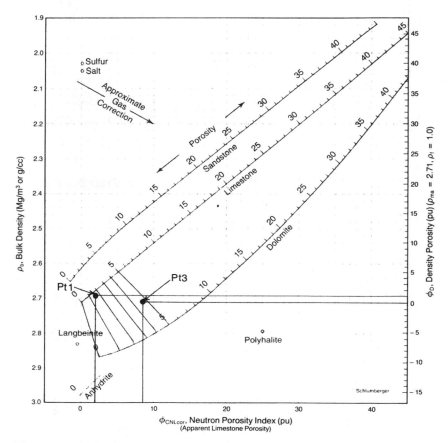

Fig. 6–12 Chart to determine neutron-density cross-plot porosity. Used with Fig. 6–11.

Now that you're started, fill out the second and third columns; then find the cross-plot porosity (ϕ_{XP}). (Answers are given at the end of the chapter.) Note that there is good agreement between the Litho-Density tool and the neutron-density cross-plot, except for point 4, because the PEFs for dolomite and shale are often similar.

To find the cross-plot porosity:

1. Make the shale corrections to ϕ_N and ϕ_D.
2. Enter the corrected density and neutron porosities ϕ_{Dcor} and ϕ_{Dcor} in the proper columns of the data sheet.

3. Enter Fig. 6–12 on the right-hand side with ϕ_{Dcor} and draw a horizontal line on the chart (at 1% density porosity for point 1).
4. Draw a vertical line through ϕ_{Ncor} (1% in this case).
5. Read the cross-plot porosity and lithology at the intersection of the two lines. (Point 1 falls exactly on the limestone matrix line at 1% porosity.)

If the intersection does not fall very close to one of the matrix lines, assume the formation is a mixture of the two matrices that the point lies between (see point 3). To help estimate the porosity, connect equal porosity points as in Fig. 6–12.

QUICK-AND-DIRTY CROSS-PLOT POROSITY

The method just explained is the most accurate way of determining the cross-plot porosity. However, there is a faster, although less accurate, way that does not use charts. To use the quick-and-dirty method:

1. Read the porosity from the logs.
2. Add the two porosities and divide by 2.
3. Correct the averaged porosity for shale with the equation:

$$\phi_{exp} = \phi_{xp} - V_{sh} \times \phi_{shxp}$$

Where:

ϕ_{exp} = effective porosity
ϕ_{xp} = averaged porosity
V_{sh} = shale volume
ϕ_{shxp} = averaged apparent porosity of shale point

Let's see how the faster method compares with the chart method. Set up a table again and calculate the cross-plot porosities for points 1 through 4. You already have a head start.

Point No.	V_{sh}, %	ϕ_D, %	ϕ_N, %	ϕ_{avg}, %	$V_{sh} \times \phi_{sh}$, %	ϕ_{xp}, p.u.
Shale	100	3.5	25	14	14	0
1	5	1	2	1.5	0.7	1*
2	0					
3	25					
4	50					

*Round off porosity to the nearest half porosity unit, such as 5.5 or 14, not 5.36 or 14.1.

Compare the three different porosity tables at the end of the chapter. There is little difference, but note that with the faster method you do not get a lithology estimate.

GAMMA-RAY LOGS

The gamma-ray (GR) log is not a porosity log, but it is usually run in conjunction with porosity logs (as well as on the resistivity log). The GR is recorded in track 1 and is used to correlate to the resistivity logs. The SP curve and the GR curve normally correlate very well because they both respond to the shale content of the formations.

The GRs that we measure with this tool are naturally occurring rays rather than induced gamma rays from a source, as in the density tool. These "natural" gamma rays emanate from radioactive potassium, thorium, and uranium, the three elements that account for most of the radiation in sedimentary formations. Potassium and thorium are closely associated with shale (illite, kaolinite, montmorillonite), while uranium may be found in sands, shales, and some carbonates.

On the whole, however, the gamma-ray curve is almost unaffected by porosity and is an excellent indicator of shale. By using the relative response of the curve compared to the 100% shale reading, we can estimate the volume of shale (V_{sh}) in the formation from the equation:

$$V_{sh} = (GR_{sh} - GR_{zone}/(GR_{sh} - GR_{clean})$$

Note the shale base line at 8 divisions in Fig. 6–13. This is GR_{sh}. It is drawn through the average reading of a thick, uniform shale. (Do not use the maximum readings; these usually correspond to high concentrations of uranium compounds.) Also note the minimum reading at point A (0.8 divisions). This is the 100% "clean" or shale-free point (GR_{clean}).

To determine V_{sh} of any zone, such as zone B, subtract the clean sand reading from the shale base-line reading. This is the denominator in the V_{sh} equation. Next, read the log at the zone for which V_{sh} is required (3.8 divisions for zone B). Subtract this reading from the shale base-line reading; this is the numerator. Divide the numerator by the denominator; the result is V_{sh}. In our example:

$$GR_{sh} = 8; GR_{clean} = 0.8 P_b = 3.8$$
$$V_{sh} = (8 - 3.8)/(8 - 0.8)$$
$$= 0.6$$

As a final review to this chapter, note Table 6–1, which lists various minerals and some of their log-derived physical parameters such as

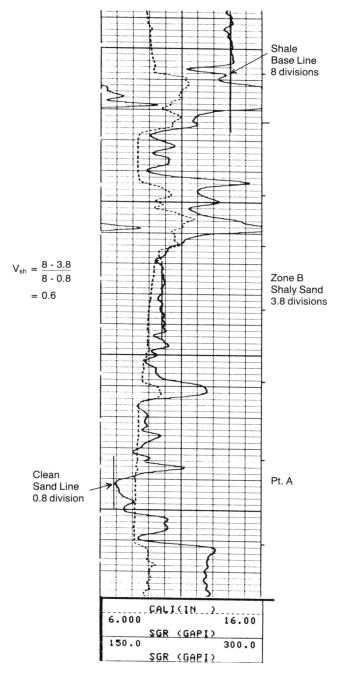

Fig. 6–13 Gamma-ray log. This example shows how to calculate shale volume, V_{sh}.

density, photoelectric index, neutron porosity, sonic compressional and shear travel times, capture cross section, dielectric constant, and natural gamma ray. Then, after you check your answers to the problems, turn to Chapter 7 so you can put all your new-found knowledge to practical use.

Answers to Problems

Density-only answers:

Point No.	Photoelectric Factor	ρ_b RHOB	$\Delta\rho$ DRHO	Porosity, %	Lithology
1	4.3	2.60	+0.01	5	Dolomitic limestone
2	4.9	2.71	+0.005	0.5	Limestone
3	4.7	2.71	0.0	1	Limestone
4	3.9	2.66	+0.07	8	50/50 limestone-dolomite

Neutron-density cross-plot answers:

Point No.	V_{sh}, %	ϕ_D, %	ϕ_N, %	ϕ_{Dcor}, %	ϕ_{Ncor}, %	ϕ_{XP}, %	Lithology
Shale	100	3.5	25	—	—	—	Shale
1	5	1	2	1	1	1	Limestone
2	0	0	2	0	2	1	Limestone
3	25	0	8.5	0	2	1	Limestone
4	50	3	19.5	1.5	7	4.5	Limestone/dolomite

"Quick-and-dirty" answers:

Point No.	V_{sh}, %	ϕ_D, %	ϕ_N, %	ϕ_{avg}, %	$V_{sh} \times \phi_{sh}$, %	ϕ_{XP}, p.u.
Shale	100	3.5	25	14	14	0
1	5	1	2	1.5	0.7	1*
2	0	0	2	1.5	0	1.5
3	25	0	8.5	4.5	3.5	1
4	50	3	19.5	11	7	4

7

PUTTING IT ALL TOGETHER

Interpreting a log is a lot like trying to hug an elephant: first you have to get your arms around it. So much data about the well is collected that it is often difficult to assimilate, or get your arms around, everything you need to make the best interpretation. Unfortunately, many decisions are made on the basis of incomplete or neglected information.

QUESTIONS TO ASK BEFORE READING THE LOG

The first step is to compile as much information as possible about the well or prospect. What is the primary objective? Is there production from this zone in the area? What type of production: oil or gas? How much? What is the cumulative production from nearby wells? Do logs exist from nearby wells? What is the lithology of the producing formation: sandstone, dolomite, or limestone? What are the porosity, resistivity, formation water resistivity, and water saturations from offset producing wells? What does the geological map look like? Are there any secondary objectives?

Answer these questions, preferably before you drill the well. Then keep the answers fresh in your mind when you examine the logs from your well. A good way to do this is to set up a table listing the various parameters from the offset wells.

In the case of a rank wildcat (a well drilled in an area that has never had a producing well), a different set of questions must be answered. These questions are based more on geologic and possibly seismic data. Estimates can be made for a best-case/worst-case scenario for porosity, resistivity, water saturation, and, ultimately, reserves. This practice usually provides the economic justification for drilling the well.

The second step is to get a feel for this well. Read the daily drilling reports. Note any unusual occurrences recorded on the mud log such as shows, kicks, lost circulation, or sticking. Next, study the mud log in detail. Again look for indications of hydrocarbons in samples that represent good formation rocks, and note signs of porosity such as drilling breaks. Talk to the mud logger if possible. He can often provide important details that are not included on the mud log.

After all this, you should have some kind of idea of what you'll see on the logs.

READING THE LOG

Normally you will have a resistivity log and one or more porosity logs to look at. One of the logs, usually the resistivity log, will have been run on a correlation scale of 1 or 2 in./100 ft. In addition, the logs will have a detail scale of 5 in./100 ft. So first compare the correlation scale with the offset wells and make them match depthwise with the SP, gamma-ray, and resistivity or porosity curves. Mark the formation tops as you identify them. After identifying the main objective and the secondary objectives, if any, mark the correlation on all the logs. Ask yourself, "Are the formations in this well running higher or lower than expected? Are they higher or lower than the offsets?" Unless a fault has been crossed, you want the well to run high because oil and gas are lighter than water and are found at the top of a zone.

Second, compare the porosity of the new well to the porosity of the offset well. Is it greater or less than the offset? How many feet of formation (net pay) are there? Does the new well have more or less reservoir than the offset? Is the reservoir quality as good as in the offset?

A good way to quantify reservoir quality is either to add the porosity foot by foot through the pay zone or to multiply the zone thickness (h) by the average cross-plot porosity corrected for shale (ϕ_e). The result is called cumulative porosity-feet. Usually a minimum porosity, called the porosity cutoff, is established; any formation with a porosity lower than the cutoff is not counted as net pay.

Third, ask yourself whether the resistivity is higher, lower, or equal to the offsets. Do any characteristic resistivity profiles suggest a gas-oil, gas-water, or oil-water contact? Is there anything on the mud log or in the known geology of the area that might suggest low-resistivity pay zones (resistivity curves affected by a low-resistivity mineral like pyrite)? Does the R_w calculated from the SP curve agree with the R_w used on the offset wells?

By examining the new well in this manner, you can decide whether the prospect should be better, worse, or about the same as the offsets. This kind of interpretive procedure takes a global approach. It assumes that nothing exists independently—that the formations are continuous and more or less alike in a given area and age. The approach is based on common sense.

Putting It All Together   117

What we've done so far is not a complete interpretation but a quick-look evaluation. It works very well when the new well has offset production. We need little expertise in log interpretation for this kind of evaluation. Naturally, more detailed methods of log interpretation are used even on development wells and must be used on wildcats where no offset production exists.

BVW$_{min}$ QUICK-LOOK METHOD

Another quick-look interpretation procedure uses the concept of minimum bulk volume water (BVW$_{min}$). This theory states that the amount of water a formation can retain without producing any water is constant for a particular formation. Therefore if we calculate BVW for a formation and it is less than or equal to BVW$_{min}$, the formation will produce water free. (Water-free production is obviously desirable because it costs money to produce and dispose of the unwanted water.) The values of BVW$_{min}$ are about 3.5% for a carbonate and from 5% for a clean sandstone to as high as 14% for a shaly sandstone.

The amount of water in the formation (BVW) is equal to water saturation times porosity (BVW = $S_w \times \phi$). Since S_w and ϕ are two of the terms in the Archie equation, let's transpose and substitute terms so we can see the significance of BVW$_{min}$:

$$S_w^2 = R_w/\phi^2 \times R_t$$

Let's see what happens if we rearrange this equation in terms of BVW:

$$S_w^2 \times \phi^2 = (S_w \times \phi)^2 = BVW^2 = R_w/R_t$$
$$BVW_{min}^2 = R_w/R_{t_{min}}$$

When terms are rearranged:

$$R_{t_{min}} = R_w/(BVW_{min})^2$$

Where:

$R_{t_{min}}$ = minimum true formation resistivity needed for water-free production

If we know the formation water resistivity and BVW$_{min}$, we can calculate the approximate $R_{t_{min}}$ needed to ensure water-free production. For a carbonate we need about 800 × R_w; for a sandstone we need about 200 (slightly shaly) to 400 (clean) times R_w. This technique, a member of the "quick-and-dirty" school of log interpretation, seems to work best in hard-rock country where matrix density (ρ_{ma}) is 2.68 for sandstones.

Once we have identified the zones that have enough resistivity to produce water-free, we only need to discover whether the porosity is high enough. For an unfractured carbonate we need at least 8% cross-plot porosity (10% for a sandstone). These are shale-corrected cross-plot porosities.

The best way to apply this technique is to draw cutoff values for resistivity and porosity on the logs. First, we determine the R_t cutoff (equal to $R_{t_{min}}$ and draw a line down the dual induction log or dual laterolog at that resistivity. Any zone with the deep resistivity curve reading to the right of the cutoff (higher resistivity) is a possible candidate for production. Next, we draw a porosity cutoff on the porosity log. Porosity cutoffs are dependent on the type of reservoir, the geologic area, and the geologist working the area. In low-porosity areas (hard-rock country), 8% for a carbonate and 10% for a sandstone are common cutoffs. In high-porosity areas, other higher values may be used. Any zones with a cross-plot porosity to the left of the porosity cutoff (higher porosity) and a resistivity to the right of the resistivity cutoff should be productive.

Normally we use either the cross-plot porosity or the density porosity. (Make any necessary adjustments for changes in lithology if you use only the density porosity.) We also note any gas effects on the neutron-density porosity log. (**Caution:** Be sure any separation on the density-neutron log is not due to a lithology change. If in doubt, see Chapter 6.)

This technique is a good scouting device that will indicate where to take a closer look, but use it with caution.

SAMPLE READING

Let's say that your Uncle Howard, the oilman, has just drilled and logged the Sargeant 1–5 well. Uncle Howard is giving you a once-in-a-lifetime chance to invest in (he says) a sure thing. To prove that he has always liked you best, he will let you see the logs before you decide whether to invest your life savings in this sure thing. Unc will give you all the information he has, but you have to make up your own mind.

On the basis of what you've learned so far about interpretation, start collecting the information you need to make your decision. Fig. 7–1 is a production map for nearby wells. You can see a 10-bcf (billion cu ft of gas) well (the Nora #1) about a mile to the north. The Corporal #1, one-half mile to the northeast, had initial production (IP) of 2 MMcfd (million cu ft of gas/day) and reserves of 2 bcf. The Private 1–5 has a

Putting It All Together 119

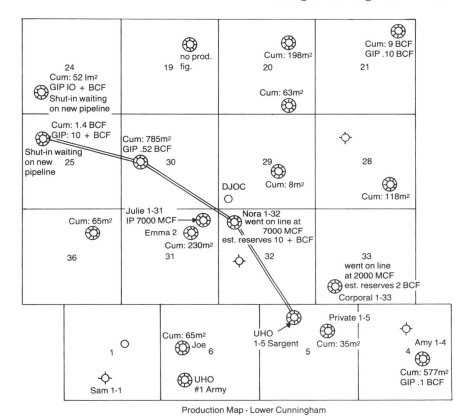

Fig. 7–1 Production map for lower Cunningham showing offset wells and their production.

cumulative production of 35 MMcf. (Unfortunately, you don't have a reserve estimate or IP.) Other apparently productive wells lie to the west and northwest. The nearest dry hole is the Sam 1–1, nearly 2 miles to the west-southwest. So far, things look good.

Now look at Fig. 7–2, the net isopach or geologic map. It shows the net zone thickness using an 8% porosity cutoff. That is, the geologist counted only that part of the reservoir that was equal to or greater than 8% porosity. He then drew his isopach (equal thickness) map.

The zone mapped is the lower Cunningham, one of the Springer series of sandstones. Each contour line represents a change in thickness of 5 ft. Zone thickness increases from 0 at the edge to a maximum of 20 ft. Note that the Nora #1 has 18 ft of pay out of a gross interval of 39 ft (18/39). The Julie that also came in at 7 MMcfd has 7 ft of net pay.

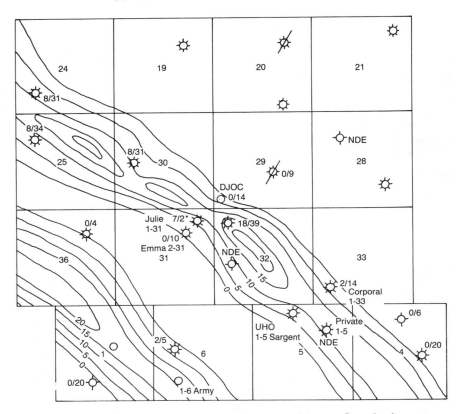

Fig. 7–2 Net isopach, or geologic map, of lower Cunningham.

The Emma has 0 ft net with 10 ft of gross pay and has produced 230 MMcf during an unknown period. Check the net and gross feet of pay on the surrounding wells. Note that your prospect, the Sargeant 1–5, is estimated to have 6 ft of net.

You're starting to amass a lot of information, so this would be a good time to set up a table to keep track of it.

Well Name	h	Reserves	IP	Cumulative Production
Nora	18	10	7	–
Corporal	2	2	2	–
Private	11	–	–	35
Julie	7	–	7	–
Emma	0	–	–	230
Sargeant	6	?	?	–

Putting It All Together 121

Table 7-1 Data on Lower Cunningham and Morrow

Well	h	ϕ	h·ϕ	R_w	R_t	S_w	BVW	Remarks
Lower Cunningham								
Nora	18	20	216	0.12	75	20	0.040	10 bcf, 7MM IP
Corp'l	2	15	32	0.12	68	28	0.042	2 bcf, 2MM IP
Emma	0	7	0	0.14	52	68	0.048	230MM cum
Julie	7	–	–	–	65	–	–	7MM IP
Private Sarg'nt	11	–	–	0.12	50	–	–	35 MM cum
Morrow								
Army	26	12	390	0.22	60	50	0.06	250M/ 150 BWPD
Private	22	10	220	0.2	42	72	0.07	not tested

As you sift through the information that Uncle Howard furnished, you find Table 7-1:

Next look at the mud log (Plate 1). Was there a show when the driller cut through the Morrow or the lower Cunningham? The formation tops have been marked on the log:

Morrow—slight show (135 units on the hot-wire gas detector; 160 units C_1 on the gas chromatograph). Background gas did not increase after drilling through the zone.

Lower Cunningham—from 12,732 to 12,755 ft the mud log shows 445 units on the hot-wire and 300 units on the GC. The formation is described as sandstone, white to light gray, fine grained, friable, subangular, fair sorting, slightly calcareous, no fluorescence, faint dull yellow ring cut. Below 12,755 ft, background gas has increased from 60 to 120 units.

From the mud log we can conclude that the lower Cunningham is promising. The results are not conclusive, however, for either zone.

Even before you look at the wireline logs on the Sargeant, you can set up some criteria on which to base your evaluation. Since the lower Cunningham is a sandstone reservoir, you'd like to have at least 10% porosity. However, the Emma has made 230 MMcfd with only 7% and the geologist mapped the formations on an 8% porosity cutoff, so use 8% as your minimum porosity. For a slightly shaly sandstone, R_t should be at least 200 times the formation water resistivity (R_w) to produce water-free. According to Table 7–1, R_w is about 0.12, so you need 200 × 0.12 = 24 ohms in the lower Cunningham. (We use the lowest R_w from the offsets because we don't want to eliminate any potential production in our scouting method. We'll refine the zones later.)

How about the Morrow? What should you use for a porosity and resistivity cutoff there? Notice that the R_w in the Morrow seems to be much fresher (higher resistivity) than in the lower Cunningham. (Check the dual induction log in Plate 2 and the compensated neutron-density log in Plate 3 where the cutoffs are already drawn.)

Since the lower Cunningham is the primary objective, look at it first. Draw the resistivity cutoff at 24 ohms on the dual induction log from the top of the lower Cunningham to the bottom of the log. To make the zones stand out, many engineers and geologists color the separation between the deep induction curve and the resistivity cutoff. Here you can see four intervals where the resistivity is higher than the cutoff, labeled 1–4.

Transfer the intervals identified on the resistivity log to the porosity log. Draw a small rectangle at each zone, 1–4. Next draw the cutoff at 8% porosity through the primary objective. Now you must approximate a cross-plot porosity calculation. Use the average of the neutron and density porosity curves, and mark this porosity at each interval, 1–4. (The average is generally very close to the cross-plot porosity that is obtained from the neutron-density chart as long as the zone is not too shaly. The average is fine for a first approximation.) Darken any of the rectangles whose porosity is 8% or greater. Only zone 2 (and not all of it) has a cross-plot porosity high enough to be darkened. Zone 1 is about 4.5%, zone 3 is 6.5%, and zone 4 is 4%. Zone 2 has 8.5% porosity and should be productive according to the quick-and-dirty technique.

Now look at the Morrow zone. We'll use the same approach. Remember R_w is at least 0.2, so draw the resistivity cutoff at 40 ohms (200 × 0.2). The lower Morrow is divided into intervals 5, 6, and 7 for easier study. The interval above the shale break is zone 8. (These zones are marked on the log in Plate 1.) Transfer the zones from the resistivity log to the porosity log and then draw the cross-plot porosity on the log (you can draw ϕ_{xp} by eye accurately enough for this step, or add ϕ_n to ϕ_d and divide by 2). Only part of interval 8 has a cross-plot porosity greater than 8%. It should also be productive.

What do you think? Will you invest the kids' college fund in this one?

Don't be hasty. The quick-and-dirty technique is just a scouting device; it points out the zones of interest so you can look at them in more detail. Go back to the table that listed some of the parameters from the offset wells and see how the Sargeant stacks up. Add what you can.

Lower Cunningham

Well	h	φ	h·φ	R_w	R_t	S_w	BVW	Remarks
Nora	18	20	216	0.12	75	20	0.04	10 bcf, 7MM IP
Corp'l	2	15	32	0.12	68	28	0.042	2 bcf, 2MM IP
Emma	0	7	0	0.14	52	68	0.048	230MM cum
Julie	7	–	–	–	65	–	–	7MM IP
Private	11	–	–	0.12	50	–	–	35 MM cum
Sarg'nt	7	8.5	60	–	60	–	–	probably gas tight

You see that the porosity is low but the porosity-feet are twice as high as the Corporal, which has reserves of 2 bcf. Also, the Emma has production with only 7% porosity. The Corporal had an IP of 2 MMcfd with 15% porosity. Since you have only 8.5%, the permeability (the ease with which the well will flow) is likely to be much less.

Before you make a final decision about whether to put your money into this well, estimate the gas reserves. You can make a rough estimate by using BVW_{min} once more. The equation for calculating gas reserves (G_p) is:

G_p = Porosity × Gas Saturation × Thickness × Area × (Initial Gas Expansion Factor − Final Gas Expansion Factor) × Conversion Constant

$= \phi \times (1 - S_w) \times h \times A \times (1/B_{gi} - 1/B_{gf}) \times 43{,}560$ cu ft

Where:

$(1 - S_w)$ = gas saturation, S_g
A = drainage area of reservoir, acres
h = ft
$1/B_{gi}$ = initial gas expansion factor = 275
$1/B_{gf}$ = final gas expansion factor = 50
conversion constant = 43,560 ft²/acre

Let:

$$1/B_g = 1/B_{gi} - 1/B_{gf}$$

Rearranging and substituting BVW for $S_w \times \phi$, you get:

$$G_p = (\phi - BVW) \times h \times A \times 1/B_g \times 43{,}560$$

Zone 2:

$$G_p = (0.085 - 0.05) \times 7 \times 640 \times 225 \times 43{,}560$$
$$= 1.54 \text{ bcf}$$

Zone 3:

$$G_p = (0.065 - 0.05) \times 5 \times 640 \times 225 \times 43{,}560$$
$$= 0.47 \text{ bcf}$$

The estimated reserves are 2.01 bcf for the lower Cunningham, your primary objective. The probable gas price is $1.25/Mcf, so you could anticipate a total sales price of $1.25 × 2,010,000 Mcf (note that 1 Mcf = 1,000 cubic feet) = $2,512,500. The net revenue interest is 78% (that is, of $1.00 in sales, 22¢ go to royalty owners and override interests) less 7.085% severance tax. So the net income will be $1,782,000. The cost of drilling and completing the well will be $1,200,000.

If you assume that the Morrow is productive, you have a G_p of 2.9 bcf. Net income from the Morrow could be as high as $2,600,000. However, there is no Morrow production in the area, so you face a good chance of no income from the Morrow. This is the way things look:

	Best Case	Most Likely	Worst
Well cost	($1,200,000)	(1,200,000)	(1,200,000)
Net income LC	1,782,000	1,782,000	1,215,000
Net income Morrow	2,600,000	1,300,000	–
Addt'l cost to complete Morrow	(150,000)	(150,000)	(150,000)
Profit	3,032,000	1,732,000	(135,000)
Return on investment	2.24	1.28	loss

Your uncle wants to cut you in for 1% of this deal (about $13,500). What do you say?

The most sensible route would be to thank your uncle and decline. With the amount of risk involved, you should get a better payout than a best case of 2.24. (Remember that a well takes several years to produce its gas.) The most likely case is that you might break even; you could certainly lose some or all of your money. But if you're still interested in the well, you need to fine-tune your calculations so that you're using the best numbers possible for your decision.

In Chapter 8, when you learn how to complete a detailed interpretation, we'll look again at Uncle Howard's well, and you'll see just how close you came with the quick-and-dirty technique. Naturally, if you are investing your own or someone else's money in a well, you would want a detailed analysis. Since this requires a certain amount of expertise, we recommend you seek someone who specializes in log interpretation, either someone in your own company, someone in one of the logging companies (usually their opinions are free), or a consultant.

8

DETAILED INTERPRETATION

Most log interpretation, at least in the basic sense of trying to determine whether a well will produce hydrocarbons, is involved with solving the Archie equation:

$$S_w^n = (F_R \times R_w)/R_t$$

Where:

S_h = water saturation
F_R = formation resistivity factor
R_w = formation water resistivity
R_t = true formation resistivity
n = saturation exponent

The formation resistivity factor is related to the formation porosity according to this relationship:

$$F_R = K_R/\phi^M$$

Where:

K_R = constant
m = cementation exponent; may vary from about 1.6 to 2.2

Common equations in use are the following:

$F_R = 1/\phi^2$ (hard-rock country)
$F_R = 0.62/\phi^{2.15}$ (Humble equation for high porosities)
$F_R = 0.81/\phi^2$ (high porosities)

To solve the Archie equation, you need to know formation water resistivity (R_w), cross-plot porosity corrected for shale (ϕ_e), and resistivity of the uninvaded formation (R_t). You can obtain R_w from water samples in offset wells, from calculations using porosity and the resistivity of an obviously 100% wet formation in the well, or from the SP curve.

Next you need porosity, and you can find it with one or more of the various porosity-measuring devices at your disposal. Normally you need more than one porosity tool to make a good estimate of porosity.

You can usually assume a relationship between formation resistivity factor (F) and porosity based on the expected range of porosities.

R_t is determined from one of the resistivity-measuring devices. Often the deep induction or deep laterolog curve is used as R_t without correction. If invasion is deep or the zones are thin (less than 10 ft), the deep curves must be corrected.

Once you have the above information on a foot-by-foot or zone-by-zone (if you want average values) basis, calculate water saturation, S_w. With S_w as one piece of your total information on the well, decide whether to run pipe, test further, or plug and abandon the well.

You can use (1) the porosity, water saturation, and thickness values, (2) a knowledge of the area that the well is expected to drain, and (3) the expected type of production to make a hydrocarbon-in-place estimate from the following equations:

For oil:

$$N = (1 - S_w) \times \phi \times A \times h \times B_o \times 7{,}758 \text{ bbl}$$

For gas:

$$G = (1 - S_w) \times \phi \times A \times h \times 1/B_g \times 43{,}560 \text{ cu ft}$$

Where:

N = total oil in place
G = total gas in place
A = drainage area, acres
h = net pay thickness, ft
B_o = oil shrinkage factor*
$1/B_g$ = gas formation volume factor†
$7{,}758$ = constant to convert acre-ft to bbl
$43{,}560$ = constant to convert acre-ft to cu ft

To convert oil or gas in place to a reserve number (the amount that can actually be recovered), multiply by the recovery factor. For oil, the recovery factor ranges from 0.05 to 0.90; 0.4 is normal. Gas recovery depends on the abandonment pressure and the type of reservoir drive; it is usually 0.7–0.9 for a water drive. For an expansion gas drive, you can use the difference between $1/B_{gi}$ (initial) and $1/B_{ga}$ (at abandonment).

---

*Because of the gas dissolved in the oil, a barrel of oil at downhole pressure and temperature will shrink to a smaller volume at the surface temperature and atmospheric pressure.
†One cu ft of gas at reservoir pressure and temperature will expand as it is brought to surface temperature and pressure.

Detailed Interpretation   129

Let's work through a couple of log interpretations and see how to come up with the various parameters you need to solve for water saturation. Start with Uncle Howard's well from Chapter 7 and see how accurate your interpretation was.

SARGEANT 1-5 EXAMPLE

You'll be working with the dual induction log (Plate 4) and the compensated neutron-density porosity log (Plate 5). Both logs have gamma-ray curves, the dual induction has an SP curve, and the porosity log has a caliper curve, all in track 1.

Usually the first thing to do is determine or check the R_w value. Remember that you used 0.12 ohms for the lower Cunningham and 0.20 ohms for the Morrow, on the basis of information from the offsets and local knowledge. However, always verify the R_w if possible by one or both of the following methods.

First look for an obviously wet sand. If you can identify one (by its low resistivity), you can calculate R_w from the equation:

$$R_o = F_R \times R_w$$
$$R_w = R_o/F_R = R_o/(1/\phi^2)$$
$$= R_o \times \phi^2 \text{ (for low } \phi\text{)}$$

Looking over the lower Cunningham, you note that nothing is obviously wet. (The ability to recognize wet [$S_w = 100\%$] zones comes with experience. Wet zones usually have low resistivities and low GR readings.) The low resistivity readings are associated with shales; you can tell this from the high gamma-ray readings and from the porosity logs. The neutron curve reads high porosity and the density reads low porosity in the shales.

On the Morrow section, one zone at about 12,640 ft has a clean gamma-ray log and 80 ohms resistivity. The cross-plot porosity is 5.5%. To check R_w:

$$R_w = (\phi^2 \times R_o)$$
$$= 0.055^2 \times 80 = 0.24$$

This is close to your assumed value for R_w, but let's make another check.

The next choice for determining or checking R_w is the SP curve. The deflection of the SP is proportional to R_w and R_{mf} (the mud filtrate resistivity). Unfortunately there is little or no movement on the SP curve in the lower Cunningham, often the case in low-porosity, low-permeability formations. (The SP will also show no deflection opposite a permeable zone if R_{mf} and R_w are the same. If R_{mf} is less than R_w, the SP will deflect to the right toward the depth track.) The upper part of the Morrow shows a small deflection of about 0.7 divisions opposite the best porosity. Let's check R_w here.

$$SP_{mv} = -K_c \log (R_{mfe}/R_{we})$$

Where:

K_c = constant that varies with temperature
R_{mfe} = equivalent mud filtrate resistivity
R_{we} = equivalent formation water resistivity

The SP equation is correctly written in terms of the ratio of the chemical activities of two sodium chloride solutions rather than resistivities. Since the resistivity of the formation water and the mud filtrate is due to other factors than sodium chloride alone, their resistivities must be converted to equivalent resistivities. Do this by using Fig. 8–1.

To calculate R_w, you need the SP reading (0.7 div × 20 mv/div = −14 mv); R_{mf} from the log heading (1.08 @ 69° F); and BHT (bottom-hole temperature) from the log heading (213° F). Then convert R_{mf} at measured temperature to R_{mf} at bottom-hole temperature with the general equation:

$$R_2 = R_1 \times [(T_1 + 7)/(T_2 + 7)]$$
$$R_{mf} = 1.08 \, [(69 + 7)/(213 + 7)] = 1.08 \, (76/220)$$
$$= 1.08 \, (76/220) = 0.37 \text{ ohms at } 213° \text{ F}$$

Next, convert R_{mf} at BHT to R_{mfe} with Fig. 8–1. Enter the bottom scale with $R_{mf} = 0.37$. Draw a vertical line until it crosses the 200° F line. Estimate by eye where 213° F is between the 200° F and 300° F line. From this point, draw a horizontal line to the left edge and read $R_{mfe} = 0.22$.

On Fig. 8–2 find the SP reading at the bottom and draw a vertical line until it intersects the 200° F temperature line. From this intersection draw a horizontal line to the right to determine the R_{mfe}/R_{we} ratio. For SP = −14 mv and 200° F, $R_{mfe}/R_{we} = 1.4$. Continue with the nomograph and solve for R_{we} by drawing a line from the R_{mfe}/R_{we} point through the R_{mfe} point on line 2 until it crosses line 3. This intersection is R_{we}. $R_{we} = R_{mfe}/(R_{mfe}/R_{we})$ and $(R_{mfe}/R_{we}) = 1.4$, so $R_{we} = 0.22/1.4 = 0.16$.

Now you must convert R_{we} to R_w. Start at the left edge of Fig. 8–1 with 0.16. After intersecting the 213° F line and dropping down to the bottom, you should come out with an R_w of 0.2. This value checks nicely with the information from your offsets, so use it for your Morrow calculations.

English

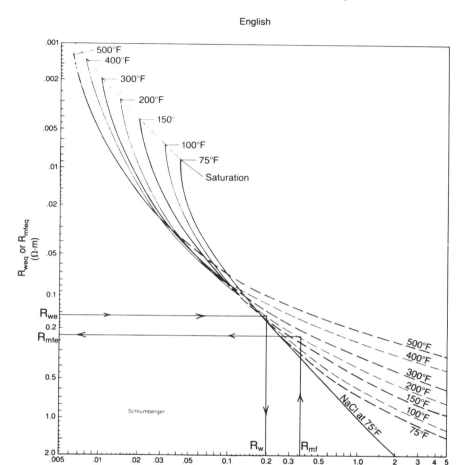

Fig. 8–1 Chart converting R_w and R_{mf} to R_{we} and R_{mfe}. The letter e stands for equivalent. (courtesy Schlumberger)

You weren't able to calculate an R_w for the lower Cunningham, so use 0.12 because this is the best information available. You were able to verify that $R_w = 0.2$ in the Morrow from the SP.

The next step is to set up a log analysis table, as shown in Table 8–1. The Sargeant has already been analyzed. Note that the table is divided into two sections, one for the lower Cunningham and one for the Morrow. Also note that the first point is a shale zone. You need the shale information to determine V_{sh} and to correct the cross-plot porosity to effective porosity.

132 Well Logging for the Nontechnical Person

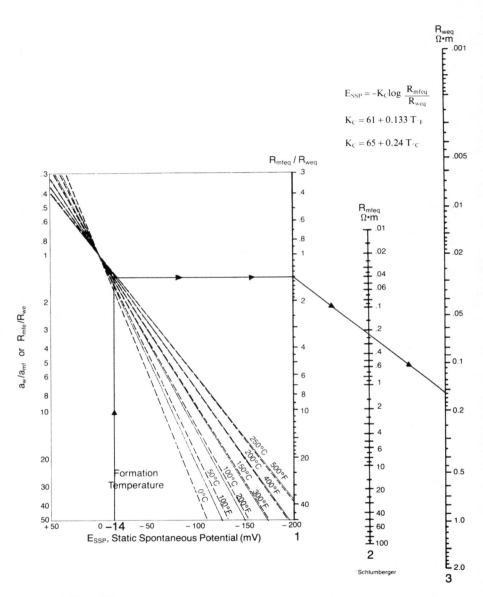

Fig. 8–2 Nomograph for determining R_{we} from the SP. (courtesy Schlumberger)

Detailed Interpretation   133

When you're working with low-porosity formations, porosity is usually the most critical input. A zone can't produce anything—even water—if it doesn't have sufficient porosity. Therefore, look at the porosity logs first and record gamma-ray, ϕ_D, and ϕ_N for each zone. When using average values for a zone, as is done here, you can see that the picks are subjective. Someone else might use slightly different numbers. Try to pick an optimistic rather than a pessimistic value. In that way, if the logs condemn a zone as wet or tight, you can be sure it is.

The next step is to calculate V_{sh}. Note the shale base line drawn on the porosity log. The lower Cunningham base line is at 6.0 divisions on track 1. The clean sand line is drawn through the lowest gamma-ray reading. GR_{cl} is usually about one division, as in this case. Now calculate V_{sh} in the lower Cunningham for zones 1 and 2:

Zone 1:
$$V_{sh} = (GR_{z1} - GR_{cl})/(GR_{sh} - GR_{cl})$$
$$V_{sh} = (1.8 - 1.0)/(6.0 - 1.0) = 0.8/5.0 = 0.16$$

Zone 2:
$$V_{sh} = (1.0 - 1.0)/(6.0 - 1.0) = 0/5 = 0$$

Where:

GR_{z1} = gamma-ray reading, zone 1
GR_{cl} = gamma-ray reading, clean
GR_{sh} = gamma-ray reading, shale

When you come to the Morrow zone, check the shale and clean sand picks. Note that GR_{sh} is now 5.0 but GR_{cl} is still 1.0 in the Morrow.

Next find the cross-plot porosity and effective porosity for each zone. On Fig. 8–3, find the density porosity on the right-hand scale and the neutron porosity on the bottom scale. Notice that the shale point ($\phi_D = 16.2$, $\phi_N = 30$) plots on the dolomite matrix line. This is where shale plots; it does not mean the zone is dolomite. The cross-plot porosity for the lower Cunningham shale is 23%.

The cross-plot porosity for zone 1 is 5.2 ($\phi_D = 7.5$, $\phi_N = 3.0$). Since some shale is in the formation as evidenced by $V_{sh} = 0.16$, we must correct cross-plot porosity (ϕ_{xp}) to get effective porosity (ϕ_e):

Zone 1:
$$\phi_e = \phi_{xp} - (V_{sh} \times \phi_{shxp})$$
$$\phi_e = 5.2 - (0.16 \times 23) = 5.2 - 3.68 = 1.52 = 2$$

Zone 2:
$$\phi_{xp} = 9.0 \; (\phi_D = 12.0, \phi_N = 5.0)$$
$$\phi_e = \phi_{xp} \text{ since } V_{sh} = 0$$

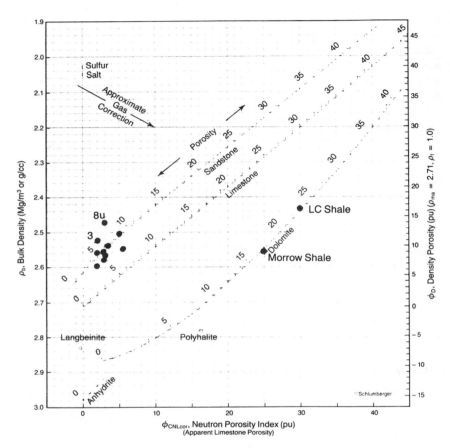

Fig. 8–3 Cross-plot porosity chart for neutron-density log. (courtesy Schlumberger)

After calculating all of the V_{sh}'s and ϕ_e's, we determine R_t for the zones. If the porosity is below 5%, we can skip the zone as being too tight to produce (unless it is naturally fractured).

Now enter the resistivity values for the shallow, medium, and deep curves in our table. If the formation is less than about 30 ft thick, the resistivity values will have to be corrected for the effect of the surrounding shales. The corrected resistivity values are then entered in the proper tornado chart where we determine R_t.

The R_t values are already entered on Table 8–1. If you must make

Detailed Interpretation   135

this kind of detailed correction, you will need a chart book from the appropriate logging company with their correction charts. Instructions for use are included in the chart books.

You can make one bed thickness correction that does not require correction charts: the $R_{t_{min}}$ method. If you wish to use this method, the bed thickness should be 10 ft to about 25 ft and the deep resistivity should be at least 10 times the shale resistivity. To find $R_{t_{min}}$:

$$R_{t_{min}} = R_s \times R_w/R_{mf}$$

Where:

R_s = shallow resistivity reading (LL8 on SFL)

This method will give a zone's minimum theoretical resistivity based on the shallow resistivity reading and the invaded zone. You can apply this method to the lower Cunningham. If $R_w = 0.12$ and $R_{mf} = 0.37$, then:

Zone 1:

$$R_{t_{min}} = 1,200 \times 0.12/0.37 = 380$$

Zone 2:

$$R_{t_{min}} = 700 \times 0.12/0.37 = 220$$

For the Morrow the zone is about 50 ft thick, so bed thickness corrections are not necessary. The R_{SFL}, R_{IM}, and R_{ID} values taken from the resistivity log are entered in the tornado chart (Fig. 8–4), and R_t is calculated from the ratio R_t/R_d. Now calculate zone 5 (12,626–12,634).

To use the tornado chart in Fig. 8–4, you need the ratios R_{SFL}/R_{ID} and R_{IM}/R_{ID}. From Table 8–1 you have $R_{SFL} = 900$, $R_{IM} = 180$, and $R_{ID} = 130$. Then:

$$R_{SFL}/R_{ID} = 900/130 = 6.92$$
$$R_{IM}/R_{ID} = 180/130 = 1.38$$

From the chart:

$$R_t/R_{ID} = 0.90$$
$$R_t = (R_t/R_{ID} \times R_{ID})$$
$$= 0.90 \times 130 = 120$$

Now that you have R_t, R_w, and ϕ_e, you can calculate S_w using the Archie equation. Finally multiply S_w by ϕ_e to find bulk volume water (BVW) and enter these two values for each zone in Table 8–1. Start with the lower Cunningham.

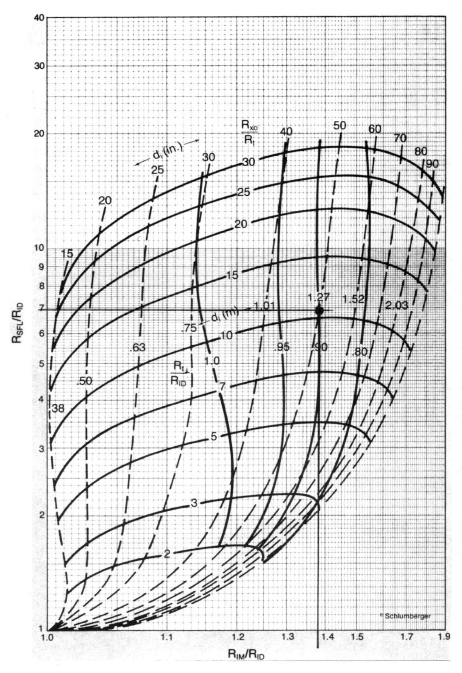

Fig. 8–4 Tornado chart for dual induction-spherically focused log. (courtesy Schlumberger)

Zone 1:
$$\phi_e = 2.0$$
Zone 1 is too tight to produce.

Zone 2:
$$R_t = 220$$
$$\phi_e =$$
$$S_w = \sqrt{R_w/\phi_2 \cdot R_t}$$
$$= \sqrt{0.12/0.09^2 \times 220} = \sqrt{0.12/1.782}$$
$$= \sqrt{0.0673} = 0.26 = 26\%$$
$$\text{BVW} = S_w \times \phi_e = 0.26 \times 0.09 = 0.023$$

Zone 2 will produce gas and no water.

Zone 3:
$$R_t = 480$$
$$\phi_e =$$
$$S_w = \sqrt{0.12/0.065^2 \times 480}$$
$$= \sqrt{0.12/2.028}$$
$$= 0.24 = 24\%$$
$$\text{BVW} = 0.24 \times 0.065 = 0.016$$

If zone 3 produces anything, it will be gas with no water, but the porosity is very low.

Zone 4:
$$\phi_e = 0$$
No production.

In the Morrow, only zone 8 appears to be productive. Here S_w and BVW are both good. The zone should produce water-free gas.

The only task left is to complete a reserve estimate to see whether enough hydrocarbons are present to complete a commercial well. The equation for gas reserves (G_p) is:

$$G_p = 43{,}560 \times (1 - S_w) \times \phi_e \times h \times A \times (1/B_{gi} - 1/B_{ga})$$

Where:

$1/B_{gi} = $ initial $1/B_g$
$1/B_{ga} = 1/B_g$ at abandonment pressure
(assume $1/B_{gi} - 1/B_{ga} = 225$)

138 Well Logging for the Nontechnical Person

Table 8–1 Data Logged

COMPANY UHOC													WELL Sargeant 1–5	
FIELD Evan Ranch							COUNTY						STATE	

DEPTH	GR	V_{sh}	R_S	R_M	R_D	$R_{T_{min}}$	ϕ_p	ϕ_N	ϕ_{XP}	ϕ_e	% POROSITY	% WATER BVW	REMARKS
12734– 12748	6.0	1.0	3	2	2	2	16.2	30	23	–	–	–	LiCl shale
12774– 12784	1.8	0.16	1200	70	52	380	7.5	3.0	5.2	2.0	88	.018	Zn1 tight
12754– 12761	1.0	0	700	110	85	220	11.0	5.0	9.0	9.0	26	.023	Zn2 gas
12739– 12744	1.1	0.02	1500	350	85	480	6.5	2.0	7.0	6.5	24	.016	Zn3 gas corrected
12694– 12698	3.5	0.50	400	40	40	130		2.0	4.0	0	–	–	Zn4

Lower Cunningham–$R_w = 0.12$ from offsets

Detailed Interpretation   139

Depth													Morrow shale Zn
12650–12660	5.0	1.0	4.5	3	3	3	9.0	25	17	–	–	–	
12640	1.0	0	380	90	80	80	9.0	2.0	5.5	5.5	91	0.05	
12626–12634	1.8	0.2	900	180	130	120	9.5	5.5	7.5	4.0	100	0.04	Zn5
12616–12622	2.3	0.32	200	90	160	160	8.5	3.0	6.0	.5	–		Zn6 tight
12604–12611	1.7	0.17	300	450	120	120	9.0	3.0	6.0	3.0	100	–	Zn7
12592–12598	1.0	0	350	120	200	200	10.0	3.5	7.0	7.0	45	.03	Zn8L
12583–12592	1.0	0	200	90	90	90	14.0	3.0	9.0	9.0	52	.047	Zn8U gas corrected

Morrow–R_w = 0.20 from SP

Subjection picks–better–every 2 ft.

All interpretations are opinions based on inferences from electrical or other measurements and we cannot, and do not, guarantee the accuracy or correctness of any interpretations, and we shall not, except in the case of gross or willful negligence on our part, be liable or responsible for any loss, costs, damages or expenses incurred or sustained by anyone resulting from any interpretation made by any of our officers, agents or employees. These interpretations are also subject to Clause 4 of our General Terms and Conditions as set out in our current Price Schedule

DATE	LOCATION	ENGINEER

Zone 2:

$$G_p = 43{,}560 \times (1 - 0.26) \times 0.09 \times 7 \times 640 \times 225$$
$$= 2{,}924{,}304{,}768 \text{ cu ft}$$
$$= 2.9 \text{ bcf}$$

Zone 3:

$$G_p = 43{,}560 \times (1 - 0.24) \times 0.065 \times 5 \times 640 \times 225$$
$$= 1.55 \text{ bcf}$$

Zone 8L (lower):

$$G_p = 1.45 \text{ bcf}$$

Zone 8U (upper):

$$G_p = 2.98 \text{ bcf}$$

For the lower Cunningham the total gas reserves are 4.45 bcf; for the Morrow, 4.43.

Making the same analysis that you made in Chapter 7 for a 78% net revenue interest (NRI) and a sale price of $1.25/Mcf produces the following table:

	Best Case	Most Likely	Worst
Well cost	($1,200,000)	($1,200,000)	($1,200,000)
Net income, L.C.	3,900,000	2,900,000	2,300,000
Net income, Morrow	3,880,000	1,725,000	–
Addt'l cost, Morrow	(150,000)	(150,000)	(150,000)
Profit	$6,430,000	$3,275,000	$950,000
ROI	4.76	2.42	0.70

This is a marginal well according to the guidelines for Uncle Howard's company. They like to have the prospect of a 4-to-1 return before completing, but only in the best case will the well have that good a return.

In real life this well was completed in the lower Cunningham only. It looks as if it will make about 2.5 bcf. In addition, the gas price has dropped to $1.10/Mcf. (Average price for the life of the well will be about $1.15.) If the Morrow is not completed, the ROI will be 1.69. Obviously the Morrow can be completed fairly cheaply, and this will be done in the future.

Detailed Interpretation 141

GULF COAST EXAMPLE

Now let's look at a suite of logs from the Gulf Coast region.

We have another gas well. The formation is an unconsolidated, high-porosity sandstone. The zone of interest is the upper Hull sand, a prolific producer in the area. We want to make a reserves estimate.

The logs that we have are a combination of Induction Electrolog/BHC Acoustilog (marks of Western Atlas*) with SP and caliper curve (Plate 6) and compensated neutron-density porosity log with gamma-ray and caliper curves (Plate 7). Note that the neutron and density logs were run on sandstone matrix (noted on the heading under logging data) because we are in an area that is predominantly sandstones.

Look carefully at the two logs and notice where the top of the formation is marked. There is an 8-ft difference between the two logs; one was not recorded properly. Mistakes like these must be recognized; otherwise, even more serious mistakes may be made, such as using porosity data with resistivity data from different parts of the formation.

The first thing to do is determine R_w. Look at zone A from 9860–66 on the Induction Electrolog. This zone is obviously wet because the resistivity is very low (0.4 ohms) and the SP has the maximum deflection seen on the logs. Find the equivalent point on the porosity logs at 9868–74. Read the porosity; both the neutron and the density read the same, which is appropriate in a clean, wet sand. Porosity is 28%. So:

$$R_o = F \times R_w$$
$$F_R = 0.81/\phi^2 \text{ for soft formations}$$
$$R_w = R_o \times \phi^2/0.81$$
$$= 0.40 \times 0.28^2/0.81 = 0.139 \text{ at BHT}$$

You can check this R_w by calculating R_w from the SP. In Plate 6, the shale base line and the maximum deflection at a sand (SSP) are already drawn for you. The shale base line is drawn through a uniform shale bed and extends above and below the zone of interest. The SSP is drawn through the largest deflection that is noted opposite a sand. In this case the maximum deflection of -5.7 divisions is at zone A.

To calculate R_w from the SP you need SP deflection in millivolts ($-5.7 \times 10 = -57$), R_{mf} at measured temperature (0.38 at 85° F), and

---

*Formerly Dresser Atlas.

bottom-hole temperature (188° F from the heading). First you must convert R_{mf} at 85° F to BHT:

$$R_{mf} \text{ @ BHT} = 0.38 \times (85 + 7)/(188 + 7) = 0.18$$

Next, find R_{mf} on Fig. 8–1 and convert it to R_{mfe} (0.14). With R_{mfe} and SP = –57 mV, enter Fig. 8–2 and determine R_{mfe}/R_{we} = 4.8. By continuing through the chart, you'll come out with R_{we} = 0.027 ohms.

Now, go back to Fig. 8–1, this time with R_{we} on the left edge, and find R_w = 0.037. This value compares well with R_w by the R_o method. Compromise and use 0.038 for R_w.

Now set up another log analysis table (Table 8–2). The listings will be slightly different because there is an additional curve, the sonic t. Since the porosity is high, the invasion will be shallow and the induction resistivity will be very close to R_t. The sonic porosity will be too high because of gas and shale effects, but it gives a good "quick look" using the R_{wa} curve in track 1 of Plate 6.

R_{wa} is the apparent formation water resistivity. Remember from the Archie equation:

$$R_o = F_R \times R_w \text{ or}$$
$$R_w = R_o/F_R$$

If you assume that every zone is wet and calculate R_w, you will get a range of R_w's. The lowest values will correspond to the wet zones, and higher R_w's or apparent R_w's will correspond to oil or gas zones. By plotting these R_{wa}'s on the log, you will see a high apparent R_w value that shows you where to make your detailed calculations. You can make this calculation because the induction log was run in combination with a sonic log. The sonic t is converted to porosity and then to F_R. The induction resistivity is divided by F_R; the output is R_{wa}, plotted in track 1.

The scale for R_w on Plate 6 is 0 to 1.0. Note how it is around 0.03 in the wet sands and then jumps to very high values through the zone, from 9,844 to 9,806. This is an indication that the zone is probably productive.

Use the gamma-ray readings for V_{sh}. GR_{sh} = 6.0 divisions and GR_{clean} = 2.2 divisions (see the lines drawn on Plate 7).

For the detailed analysis, use the cross-plot porosity from the neutron-density log. Although there are charts that help us to find ϕ_{xp}, it is just as accurate and much quicker to use the equation:

$$\phi_{xp} = (1.5 \times \phi_d + \phi_n)/2.5$$

This relationship should be used in gas zones with reservoir pressures less than about 5,000 psi. The density porosity is weighted because it is less affected by the gas than is the neutron porosity.

In high-pressure gas zones the following equation should be used:

$$\phi_{xp} = (\phi_D + \phi_N)/2$$

ϕ_e is calculated from the relationship:

$$\phi_e = \phi_{xp} - (V_{sh} \times \phi_{shxp})$$

The next column in Table 8–2 contains a new parameter: porosity-feet. This parameter is simply the effective porosity times the depth interval between calculations. Here you are making a calculation every 4 ft (in reality you'd make one every 2 ft), so multiply $\phi_e \times 4 = \phi$-ft. To find the total porosity-feet for the zone, just add the ϕ-ft column (8.66). If you want the average porosity for the zone, divide by the zone thickness (44 ft) so that $\phi_{avg} = 8.66/44 = 0.197 = 19.7\%$.

Next calculate the water saturation for each point. The presence of shale in the formation reduces the true resistivity. If you used the Archie equation for S_w, you would get values that are too high and you might miss a productive zone. Many different methods have been developed to calculate shaly sand water saturations; the method you'll use here was developed by Simandoux and is relatively straightforward:

$$S_w = \frac{c \times R_w}{\phi^2} \left[\sqrt{\frac{5 \phi^2}{R_w \cdot R_t} + \left(\frac{V_{sh}}{R_{sh}}\right)^2} - \left(\frac{V_{sh}}{R_{sh}}\right) \right]$$

Where:

c = 0.4 for sands and 0.45 for carbonates
V_{sh} = shale volume
R_w = formation water resistivity
R_t = deep resistivity in zone of interest
R_{sh} = deep resistivity in nearby shale
ϕ_e = effective porosity

Note that when $V_{sh} = 0$, you can use the simpler Archie equation. Although the Simandoux equation is cumbersome in hand calculations, even a simple programmable calculator can handle it easily.

Now calculate S_w using R_{ID} for R_t, $R_w = 0.038$, and $R_{sh} = 0.9$ with the values for V_{sh} and ϕ_e from Table 8–2, and list your results in the S_w column of the table.

Now we come to another new parameter: hydrocarbon-feet (hc-ft). This is simply porosity-feet times $(1 - S_w)$. (Remember that the hydrocarbon saturation $S_h = 1 - S_w$.) From hc-ft you can find the av-

144 Well Logging for the Nontechnical Person

Table 8-2 Data Logged

COMPANY
Shaly Sand Example

FIELD
Evan Ranch

COUNTY

WELL

STATE

DEPTH	R_D	Δ_t	ϕ_s	GR	V_{sh}	ϕ_D	ϕ_N	ϕ_{xP}	ϕ_e	ϕ-ft	S_w	h_i,ft	BVW	REMARKS
9856	1.0	95	30	2.4	0.05	24	30	27	25.5	—	69	—	.175	Wet
52	.3	96	30.5	3.3	0.29	25	25	25	16.5	0.66	100	0	.165	Wet
48	.45	94	29	2.6	0.11	26	30	28	25	1.00	100	0	.25	Wet
44	2.0	96	30.5	2.2	0	30	23	27	27	1.08	46	0.58	.15	Wet
40	8.5	96	30.5	2.2	0	32.5	12	24.5	24.5	0.98	25	0.74	.06	Gas
36	10.0	84	21.5	2.2	0	34	13.5	26	26	1.04	21	0.82	.055	Gas
32	20	92	27.5	2.4	0.05	32.5	12	24.5	24.5	0.98	25	0.74	.06	Gas
28	12	84	21.5	2.7	0.13	25.5	14.5	21	17.5	0.70	22	0.55	.039	Gas

20	21	70	12	2.3	0.03	13.5	9	12	11	0.44	31	0.30	.034	Gas
16	15	82	20	2.5	0.08	11	11	11	9	0.36	36	0.23	.032	Gas
12	12	94	29	2.7	0.13	21	12	17.5	14	0.56	26	0.41	.037	Gas
08	2.6	103	36	3.0	0.21	7	12	9.5	3.5	—	—	—	—	Tight
04	6.0	60	3	2.7	0.13	0	9	4.5	1	—	—	—	—	Tight
Shale	0.9	96	30.5	6.0	1.0	23	34		Σ	8.66		5.21		
								28.5	—					

For ϕ_{XP} use: $\phi_{XP} = (1.5\,\phi_D + \phi_N)/2.5$

GR clean = 2.2 div. R_w = .038

All interpretations are opinions based on inferences from electrical or other measurements and we cannot, and do not, guarantee the accuracy or correctness of any interpretations, and we shall not, except in the case of gross or willful negligence on our part, be liable or responsible for any loss, costs, damages or expenses incurred or sustained by anyone resulting from any interpretation made by any of our officers, agents or employees. These interpretations are also subject to Clause 4 of our General Terms and Conditions as set out in our current Price Schedule.

DATE	LOCATION	ENGINEER

erage water saturation, and you can use hc-ft very conveniently when you make your reserve calculation.

The final column is bulk volume water. On inspecting the BVW column, you see that there is a gas-water contact between 9,844 and 9,840 ft. Call 9,840 the bottom of the productive zone. The top is 9,812; above that, the zone is tight.

Now let's perform a reserve calculation. This is a water drive reservoir, and the recovery factor is estimated to be 80%. Use $1/B_g = 175$ (the gas expansion factor):

$$\text{Reserves} = 43{,}560 \times [h \times \phi \times (1 - S_w)] \times A \times 1/B_g \times RF$$

The term in the brackets is the hydrocarbon-feet, so you can obtain this from Table 8–2. The total hydrocarbon-feet value from 9,840 to 9,812 is $5.21 - 0.58 = 4.63$:

$$\begin{aligned}\text{Reserves} &= 43{,}560 \times 4.63 \times 160 \times 175 \times 0.8 \\ &= 4.52 \text{ bcf}\end{aligned}$$

In a nutshell, this is how you would work through a detailed log interpretation. Today, however, computers are used to help us generate and interpret logs. To learn more, read Chapter 9.

9

COMPUTER-GENERATED LOGS

The amount of information available from well logs today is truly amazing, especially when you consider that only 20 years ago a sophisticated interpretation was a moved oil plot or an S_w calculation made by hand every 2 ft.

The introduction of new radiation detectors, downhole microprocessors (minicomputers), and especially the truck-mounted (onboard) computer has had the most profound impact on the logging industry since the first log was run. The new sensors and processors have made a multitude of measurements possible (gamma-ray spectroscopy, induced spectroscopy, Litho-Density, sonic waveform recording, stratigraphic dip measurements) that were unknown only a few years ago except in laboratories. Onboard computers have made all this possible. The computer handles the wealth of information that is transmitted by the downhole telemetry system, stores it on magnetic tape, corrects for environment, merges data, makes complex calculations, and prints out various log formats, simultaneously keeping the logs on depth and warning the logging engineer if a tool malfunctions.

Without a doubt, the computer is a wonderful labor-saving tool. It performs calculations and data presentations that would not be practical by hand. The computer's strength is its speed and ability to make repetitious calculations; its weakness is that it does only what it is told. It cannot think (at least not yet) and cannot make judgments or true interpretations. We still need humans for that.

Computer logs are generated either at the wellsite or in computing centers. They are usually a combination of two or more measurements, with some sort of calculation taking place. The main difference between wellsite and computing center logs is the size of the computer making the calculation. The more powerful the computer, the more data it can handle and the more complex the software programs and calculations it can make.

WELLSITE COMPUTER LOGS

Wellsite computer logs are often called quick-look logs. They indicate water saturation and porosity, for example, without the detailed analysis that the computing center logs make. All of the major logging companies and many of the smaller ones have onsite computing capabilities.

The most common onsite computer solution log uses deep resistivity, neutron and density porosity, gamma-ray, SP, and caliper curves to solve for water saturation. The log values are first corrected in a preliminary interpretation pass for environmental effects (temperature, borehole size and salinity, and mud weight). On this pass (Fig. 9–1), the corrected porosities from the neutron and density logs are cross-plotted. The logging engineer then picks the various parameters (R_w, GR_{sh}, SP_{sh}, $\phi_{N_{sh}}$ etc.) needed to make the final interpreted log.

Fig. 9–2 is the wellsite computer-generated log for the Sargeant 1–5. This is an excellent example of both the strengths and weaknesses of a computer-generated log. In this case an R_w of 0.09 was used throughout the lower Cunningham and the Morrow. You know both from offset production and from calculations on the logs that the R_w is closer to 0.12 from the lower Cunningham and 0.2 for the Morrow. Using 0.09 instead of the correct R_w will give water saturations that are 15% too low for the lower Cunningham and 33% too low for the Morrow. This variation is large enough to cause a major mistake, such as running pipe on a wet well or abandoning a productive well.

The major precaution when looking at someone else's interpretation—even a computer's—is to check all of the assumptions, input, and methods. Why was a particular R_w chosen? Does it agree with what was used in other wells in the same zone? What was used for the shale values? Were log values corrected for environmental effects? These questions are important because they affect the corrections made to the log readings in the zones of interest. Was the well properly zoned? In other words, were the necessary parameters changed as the interpretation was made in a different formation?

Obviously the major strength of computer-generated interpretation logs is that calculations are made continuously; all we have to do is read values of ϕ_a, S_w, and BVW directly from the log (after we have checked the assumed values). The computer helpfully shades the area between BVW and ϕ_e so that the amount of hydrocarbon stands out, or it indicates different rock types by changing the shading or dot pattern between curves.

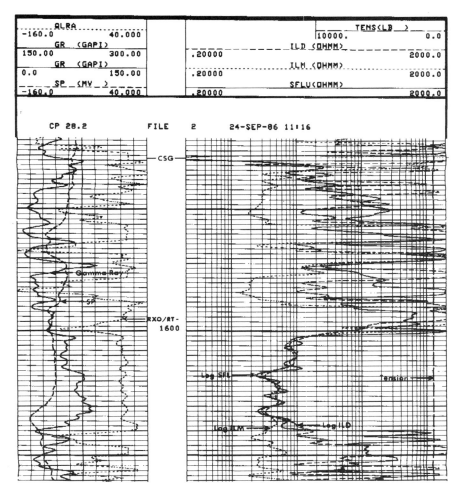

Fig. 9–1 Wellsite pre-interpretation pass, used to make environmental corrections and to pick the parameters used in the interpretation pass.

KINDS OF WELLSITE-COMPUTED LOGS

Wellsite logs come in many different formats. The drilling engineer needs different information from the geologist, for example, so wellsite computed logs may be classified according to the end user—geologist, seismologist, drilling, completion, production, or reservoir engineers—and the presentations tailored to their needs.

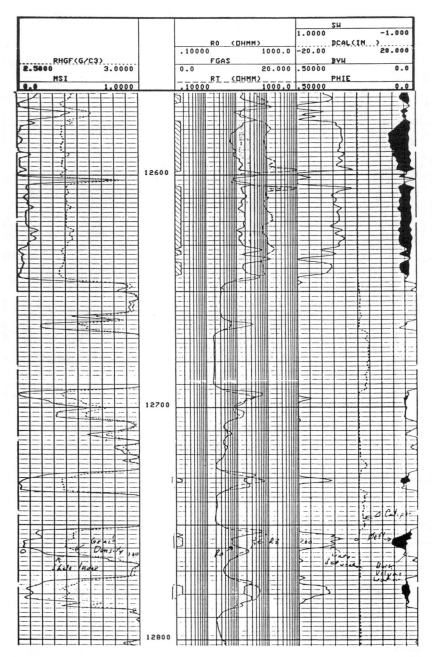

Fig. 9–2 Wellsite computer-generated log for the Sargeant 1-5. Note that the water saturations are higher than those calculated in Chapter 8. See text for explanation.

Computer-Generated Logs 151

Table 9–1 Logs and Their Uses

For the Geologist:
- Wellsite S_w, ϕ_e, and V_{sh}—used for pipe-setting decisions
- Lithology Analysis—calculates the percentage of limestone, shale, and dolomite/shale present; requires the photoelectric index
- Wellsite Dipmeter Computations—make preliminary dip calculations—used when a quick decision is necessary
- Fracture Identification—indicates presence of natural fractures
- True Vertical Depth—used when a high-angle wellbore distorts thickness and depth of formations

For the Seismologist:
- Wellsite Two-Way Travel Time—converts and corrects the data from seismic check shots
- Vertical Seismic Profiling—processes the data from geophones in the wellbore to provide a seismic record that may be correlated with the surface seismic data

For the Drilling and Completion Engineer:
- Borehole Profile and Cement Volume—used to calculate amount of cement needed (Fig. 9–3)
- Directional Log—uses data from dipmeter to calculate location of wellbore with respect to surface location
- Cement Bond Log/Cement Evaluation Log—evaluates cement job

For the Production Engineer:
- Wellsite Production Log—calculates amount of gas, oil, and/or water coming from a zone
- Wellsite S_w, ϕ, V_{sh} Through Casing—uses pulsed neutron log to evaluate zones behind casing

---

The quick-look S_w–ϕ_e presentation that you've already examined is used primarily by the geologist and management to make a pipesetting decision. If casing is run, the drilling or completion engineer will need to know how much cement to order. He gets this from the cement volume log. Specific uses of logs and how they affect interpreters are listed in Table 9–1.

All of the logs that are interpreted at the well are considered quick-look answers. Normally more detailed interpretations are made at the computing centers. The quick-look logs are just that: they are used

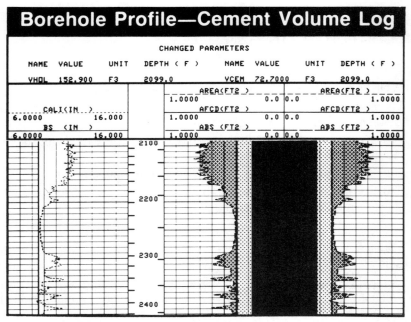
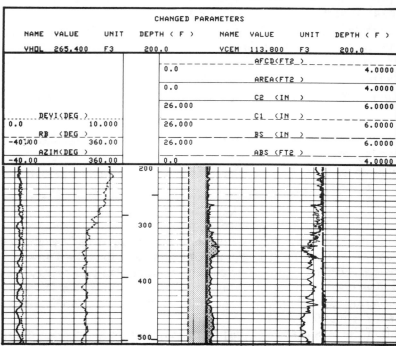

Fig. 9–3 Borehole profile and cement volume (courtesy Schlumberger). Used by the drilling engineer to determine how much cement will be needed to cement the casing.

Computer-Generated Logs 153

when immediate answers are needed. If time permits, the interpretations made by the computing centers are preferred. (This situation may be changing, however, because more powerful onboard computers are now available.)

COMPUTING CENTER LOGS

Fig. 9–4 is another wellsite computed log similar to Fig. 9–2 except for the addition of the photoelectric index curve, which indicates lithology. The lithology is shown in the depth track as either limestone (no coding), sandstone (coarse dot pattern), or dolomite or shale (fine dot pattern). Also note that track 1 includes a dashed coding for the shales. Once again the area between ϕ_e and BVW is coded black to indicate the hydrocarbon.

Compare Fig. 9–5, a log generated and interpreted at a computing center, with Fig. 9–4—same well, same interval. Immediately we notice the change in format. The depth track is on the left edge of the log. Gamma-ray and apparent grain density curves (this is a calculation based on ϕ_{xp} where a matrix density is back-calculated from ϕ_{xp}) are in track 1, as is a permeability indicator. Track 2A shows water saturation (and residual hydrocarbon volume if an R_{xo} tool was run); track 2B shows ϕ_e and BVW. The caliper curve is presented with the bit size at 0 divisions of track 2B. Track 3 shows the bulk volume analysis of the entire formation. In this case it is divided into dry clay, bound water (water associated with the dry clay), silt, matrix, and effective porosity. The porosity is further divided into water and hydrocarbon. The presentation varies, depending on the logs run.

Note that water saturations are slightly higher and porosities are lower on the computing center log. These logs are considered to be more accurate interpretations than field-generated computer logs, mainly because the more powerful computers and programs used in the computing centers allow more information to be used and more complex corrections to be applied than at the wellsite. Also, a specialist in computer logs makes the interpretation, often with input from the oil company log analyst, while the field engineer probably has less expertise.

Also included with the more advanced interpretation is a listing (Fig. 9–6) of the more important outputs (S_w, ϕ_e, permeability, grain density, and V_{sh}) as well as some additional calculations (cumulative porosity-feet and hydrocarbon-feet) that are useful for reservoir and hydrocarbon-in-place calculations. Look at the zone at 6,340–6,344 on the computed log; then find this zone in the listing. You can

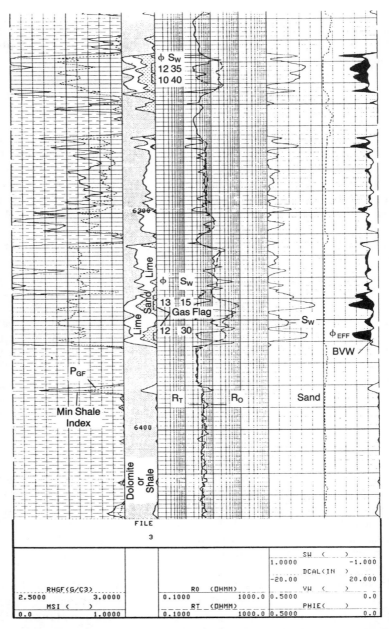

Fig. 9–4 Wellsite computer-interpreted log. Note the lithology information in the depth track.

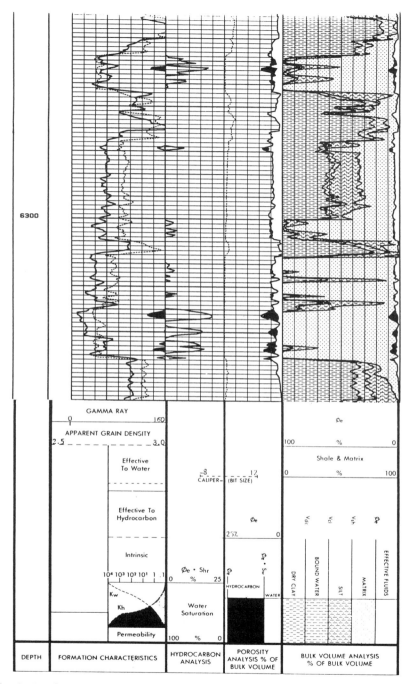

Fig. 9–5 Computing center-interpreted log. Note the different format, with the depth track on the left. This is a convention adopted by logging companies to distinguish computing center logs from wellsite logs.

COMPANY : SALES EXAMPLE WELL : PAGE 5

DEPTH FT	WAT. SAT. PU	POR. EFF. PU	POR. 2ND PU	CUM. EFF. POR. FT	CUM. HC-FT FT	PERM. INTRIN MD	PERM. GAS MD	PERM. OIL MD	PERM. WATER MD	CUM. PERM. GAS FT	CUM. PERM. OIL FT	CUM. PERM. WATER FT	CLEAN GRAIN DENS G/C3	VOL SH PU
6268	85	8.		4.1	.6	.2	0	0	0	6.	0	0	2.70	37
6269	79	8.		4.0	.5	.2	.1	0	0	6.	0	0	2.70	36
6300	100	5.		2.7	.5	0	0	0	0	6.	0	0	2.80	12
6302	87	5.		2.6	.5	0	0	0	0	6.	0	0	2.84	15
6334	66	5.		1.5	.5	0	0	0	0	6.	0	0	2.79	3
6340	67	5.		1.3	.5	0	0	0	0	6.	0	0	2.71	16
6341	42	7.		1.2	.4	.3	.3	0	0	6.	0	0	2.69	0
6342	23	9.		1.2	.4	2.5	2.5	0	0	5.	0	0	2.69	0
6343	20	8.		1.1	.3	1.9	1.9	0	0	3.	0	0	2.69	0
6344	39	5.		1.0	.3	0	0	0	0	1.	0	0	2.70	0
6349	41	5.		.9	.2	0	0	0	0	1.	0	0	2.71	0
6350	38	6.		.8	.2	.1	.1	0	0	1.	0	0	2.71	0
6355	45	7.		.6	.1	.3	.3	0	0	1.	0	0	2.69	0
6357	34	8.		.5	.1	.6	.6	0	0	1.	0	0	2.69	13
6358	52	6.		.4	0	.2	.1	0	0	0	0	0	2.70	0
6361	65	8.		.3	0	.5	.2	0	0	0	0	0	2.69	24

Fig. 9–6 Listing for Fig. 9–5. Parameters of interest are tabulated for easy use in reserve calculations.

see that S_w varies from 67% to 20% and ϕ_e from 5% to 9%. The column for cumulative porosity-feet sums up (integrates) the porosity. To find the number of porosity-feet number for the zone, simply subtract 0.9 (cumulative porosity-feet for the zone just below the one you want) from 1.3 (the total porosity-feet for the well from total depth to 6,340). The result is 0.4. Out of 4 ft of formation, 0.4 ft of space is available to hold fluids. This value can be used to draw net isopach maps (net porosity maps).

The next column is cumulative hydrocarbon-feet just, $S_w \times \phi_e$ for each foot of formation. You can use this number in your reserves calculation by subtracting the number of hydrocarbon-feet in the last depth below the zone of interest (0.3) from the number of hydrocarbon-feet at the top of our zone (0.5). (This is the term $[1 - S_w] \times \phi_e \times h$ in the reserves calculation.) To calculate the reserves for this 4-ft zone:

$$G_p = 43{,}560 \times (1 - S_w) \times \phi_e \times h \times A \times 1/B_g$$

But $(1 - S_w) \times \phi_e \times h$ = cumulative hydrocarbon-feet for the zone. So:

$$G_p = 43{,}560 \times (0.5 - 0.3) \times 160 \text{ acres} \times 125$$
$$= 0.174 \text{ bcf}$$

Computing centers can turn out more comprehensive interpretations, often with input from nonlogging sources such as core analysis, that allow insight that was scarcely dreamed of 20 years ago.

With Faciolog (mark of Schlumberger), various sediments or facies are identified by their unique log responses (Fig. 9–7). The logs required include resistivity, Litho-Density, compensated neutron, dipmeter, and core data.

Rock properties logs determine parameters such as Poisson's ratio, Young's modulus, shear modulus, and closure stress from acoustic (sonic) measurements. These logs may be used to predict the response of the formation to a hydraulic fracture treatment or to determine whether there will be problems with sand flow in unconsolidated formations.

Stratigraphic dips within sand bodies as well as structural dips can be determined with the newer measuring systems that take readings much more often. These logs include structural and stratigraphic dipmeter presentations with supplemental plots such as polar and azimuth frequency plots.

With reservoir descriptions, any data available on a multiwell reservoir, including log, core, production, and geophysical data, are used to create a data base. The data base helps us correct bad data, zone the wells, distinguish trends, or create algorithms for the reservoir.

158   Well Logging for the Nontechnical Person

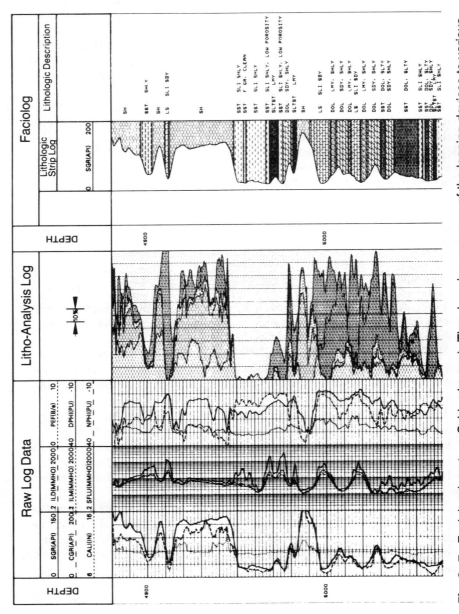

Fig. 9–7 Faciolog (courtesy Schlumberger). The changing response of the logging devices to various formations allows the identification of different facies.

The various parameters may be displayed in a number of different ways, including colored 3-D maps.

Computer-generated logs have changed the face of logging. In addition, specialty logs and tools help the oilman locate hydrocarbons trapped deep in the earth—the subject of our final chapter.

10

SPECIALTY LOGS

The logging tools discussed in earlier chapters are those most commonly run. The usual suite of logs consists of a multiple resistivity log with a gamma-ray or SP log and at least one porosity log. If gas is suspected or the lithology is unknown, two porosity devices are run. This logging suite is designed to determine porosity, water (or hydrocarbon) saturation, and reservoir quality (shaliness). These data are the very minimum information an engineer needs when deciding to set pipe.

However, geologists and engineers would like to have much more information, such as where to drill next in a channel sand reservoir. Also, techniques such as air drilling require special methods of formation evaluation. To meet some of these special needs, the following tools were developed.

DIPMETER

One of the basic tenets of geology is that sedimentary rocks or beds are deposited horizontally, usually on the bottom of a lake, sea, or ocean. Therefore, the original dip of a reservoir rock is zero. However, this condition rarely persists. Because of tectonics (forces within the earth that cause subsidence, mountain building, earthquakes, etc.), the horizontal beds are tilted, faulted, compressed, and eroded. Often the formerly horizontal beds are shaped into traps, structures that inhibit the migration of hydrocarbons toward the surface. When we can identify the structure, we can develop the oil field properly, i.e., we can choose accurate drillsite locations in optimum places.

By measuring the dip (the angle that the formation makes below the horizontal) and the azimuth (the direction in which the bed is dipping) of the formations, we can infer a great deal about the processes that moved the beds to their present positions. From dip information we can determine many factors that help engineers determine how to develop a field or where to drill an offset well.

To calculate dip, we need a large amount of information (Fig. 10–1). Remember from high school geometry that any three points not on a straight line define a plane. Since the wellbore is round, we can measure (1) the depth of a formation feature (high resistivity point,

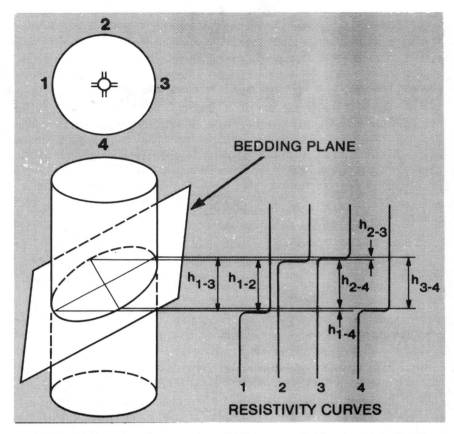

Fig. 10–1 Principle of measuring dip (courtesy Schlumberger). See how the displacements relate to the intersection of a plane through a borehole.

for example) at three different points, (2) the displacement between the three points, and (3) the wellbore diameter. Then we can calculate the apparent dip using simple geometrical relationships.

To make these measurements, four microresistivity pads (instead of three) are pressed against the wall of the borehole (Fig. 10–2). The curves generated by these pads are correlated with each other to define the displacement needs in point 2 above. A caliper curve is recorded at the same time. To determine the azimuth of the dip, we need to know which direction the three pads are facing when the measurements are made. We do this by using a magnetic compass that is aligned with pad 1. From this information, we can calculate dip angle and direction if the hole is straight.

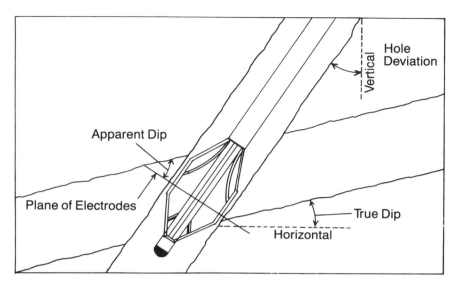

Fig. 10–2 Geometry of dip measurement (courtesy Schlumberger). Relationship between apparent dip and true dip.

Unfortunately, wellbores are seldom perfectly straight horizontally. The angle the borehole makes to the vertical is called hole deviation or hole angle. Therefore, we must measure the hole angle as well as the direction the hole is going (hole azimuth). From all of this information we can calculate the actual dip direction and angle.

In practice, four curves are used to measure the dip. There are two reasons for using more than the minimum three required. First, if all four points fall on the same plane, we can have more confidence in our calculation. Second, if one of the pads is not contacting the formation for some reason, we still have three curves that are reading correctly.

Today, dip calculations are made by computer. Computers can process a large number of calculations and allow a variety of interpretations. Relatively few points are used in the structural dipmeter presentation because structure changes slowly. For stratigraphic dips a special dipmeter tool with eight correlation curves is used; the larger number of correlation curves lets engineers calculate more dips. This is necessary for detecting the subtle changes in stratigraphy.

Fig. 10–3 is a presentation of a stratigraphic dipmeter log, often called a tadpole plot. Notice the detail in the presentation. In this kind of log we can often see cross-bedding in the sands or channel filling. The structural dipmeter, which has less detail, shows large features such as faults.

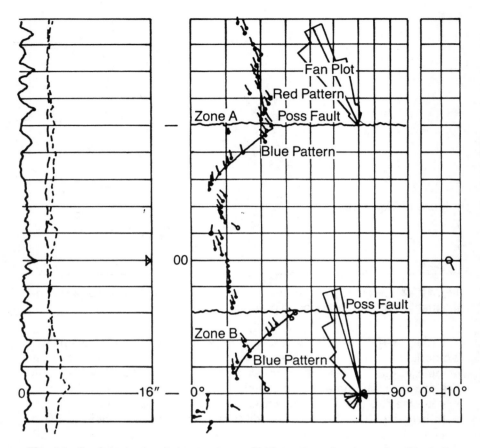

Fig. 10–3 A tadpole plot, one way of presenting dip data. Angle of dip increases to the right of the graph; arrows (tadpoles) point in the direction of dip.

By using dip information, we can discover anticlines, faults, or unconformities as well as subtleties such as sand transport or cross-bedding. The geologist may refine his maps and more confidently pick his next well location with this information.

LOGS FOR AIR-DRILLED HOLES

In some parts of the country, wells are drilled with air used as the circulating fluid rather than mud. This procedure is possible when formation pressures are low, the formation is tight, and water is not present. The main reason for drilling with air is that it is usually faster and cheaper than drilling with mud. However, the lack of a conductive mud system rules out the use of any logs that require a conductive fluid, such as microlog, SP, laterolog, or compensated neutron log.

The usual logging suite for these wells is an induction log (only one resistivity measurement is necessary because there is no invasion), a compensated density porosity log, an epithermal neutron porosity log, and often a temperature and noise log.

Epithermal Neutron Log

The epithermal neutron log is similar to the regular neutron log except that it detects high-energy epithermal neutrons. The source and detectors are mounted on a skid that is pressed against the side of the wellbore, just like a density tool. The log is used in air-drilled holes because it can measure apparent porosity in air, unlike the compensated neutron tool.

Temperature Log

The temperature log measures the changes in the temperature of the air or gas in the wellbore as the tool is lowered. Normally temperature increases with depth; however, when gas enters the hole from the formation, there is a cooling effect due to adiabatic expansion. Cooling anomalies are of interest to engineers because they indicate gas entry.

Noise Log

The noise log is essentially a microphone that amplifies any noise or sounds detected downhole. The frequency of the sound encountered is recorded. Under optimum conditions the type of fluid entering the wellbore from the formation—gas or liquid—can be determined from the frequency.

DIELECTRIC CONSTANT LOGS

Occasionally formations will be encountered whose formation water is very fresh—high in resistivity—or where the cementation exponent

(m) and the saturation exponent (n) in the Archie equation are not as assumed. In these cases, normal interpretation techniques break down. The fresh water causes high resistivity readings, just as an oil or gas zone does. In the case of freshwater formations, it is impossible to calculate a reliable water saturation using only porosity and resistivity measurements. When m or n in the Archie equation are not constants equal to 2, we cannot solve the equation for water saturation because there are too many unknown factors.

Dielectric constant logs were developed to help solve these problems. Dielectric constant is a physical parameter like density or resistivity. It is a function of the polarizability, dipolar relaxation, and conductivity of a material. The dielectric constant of fresh water is about 80; that of oil is about 2. So, as you can see, there is a strong contrast in values between fresh water and oil. This contrast gives us a way to distinguish oil from water directly.

Unfortunately, the measurement has a very shallow depth of investigation (1–3 in.). This means the measurement is being taken in the flushed zone, which is filled mainly with mud filtrate. Interpretation is complicated, but the dielectric constant logs are a definite aid in evaluating zones with fresh water or formations in which the cementation and saturation exponents are not constant.

NATURAL GAMMA-RAY SPECTROSCOPY LOG

The natural gamma-ray curve has its source essentially in three radioactive isotopes: uranium (U), thorium (Th), and potassium (K). Thorium and potassium are usually found in shales and clays. Uranium compounds may be found in practically any formation.

The development of new, more sensitive detectors and more sophisticated electronics such as multichannel analyzers has allowed the separation of the natural gamma-ray measurement into its three major components (U, Th, and K). The gamma-ray spectroscopy tool is often run in place of the standard gamma-ray tool (Fig. 10–4).

We have long used the total natural gamma-ray measurement as a shaliness indicator, but it can be misleading because uranium may be associated with both shale and reservoir rock. A better shaliness determination may be made from the thorium and potassium measurements. Engineers can also type, or identify, the clays (montmorillonite, illite, kaolinite) with these measurements because different clays have different ratios of thorium to potassium.

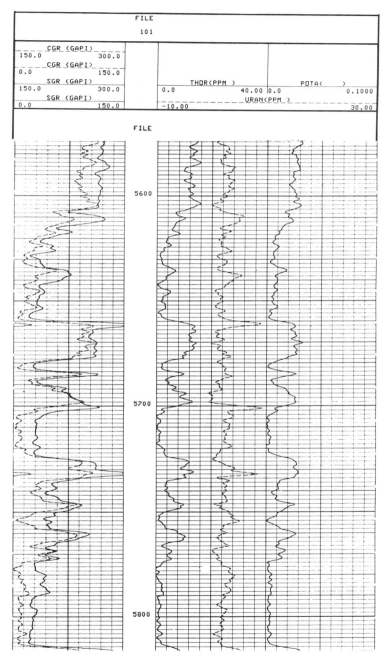

Fig. 10–4 Gamma-ray spectroscopy log. Log shows separation of thorium, potassium, and uranium.

FORMATION TESTING

Formation testing is an important source of information on formations and their fluid content. While cores are commonly used to determine or verify porosity, permeability, lithology, and fluid type and saturation, formation tests are used to determine the ability of the formation to flow, the formation pressure, and the type of fluids that the formation contains. Formation tests may be made with tools run either on the drillpipe (drillstem tester) or on an electric wireline (wireline formation tester).

Drillstem Testing

The drillstem test (DST) is performed by running test tools that consist of open-hole packers and valves (Fig. 10–5). After the packer is set above the zone to be tested, a downhole valve is opened so the well can flow to the surface through the drillpipe. The drillpipe usually contains some fluid (a cushion) but is not full.

The fluids in the formation are at reservoir pressure. When the test tools are opened to flow, fluid flows out of the rock, into the wellbore, and up the drillpipe. The downhole pressure gauge and recorder measure the pressure at all times. During the flow period the pressure measured is called flowing pressure. After a certain flow period determined by the engineer, the test tools are closed or shut in. The maximum pressure measured during this time is called the shut-in pressure. As soon as the tools are closed, the well stops flowing and the pressure begins to build up from flowing presure to reservoir pressure. Typically, the test tools are opened for a certain flow period, shut in for a period twice as long as the flow time, reopened, then shut in again.

If fluid reaches the surface, it is routed to a separator where the gas is metered and the liquids are measured and sent to a tank. If the well does not flow to the surface, the driller will be able to note any rise in fluid level when the pipestring is pulled. He can measure the level by counting the pipes that have been stacked in the derrick.

The chart or log of pressure vs. time is called a pressure buildup curve (Fig. 10–6). Analysis of this curve helps engineers calculate reservoir permeability and pressure as well as formation damage.

DSTs often take a long time to perform and so use up much valuable rig time. They are also expensive, especially compared with coring.

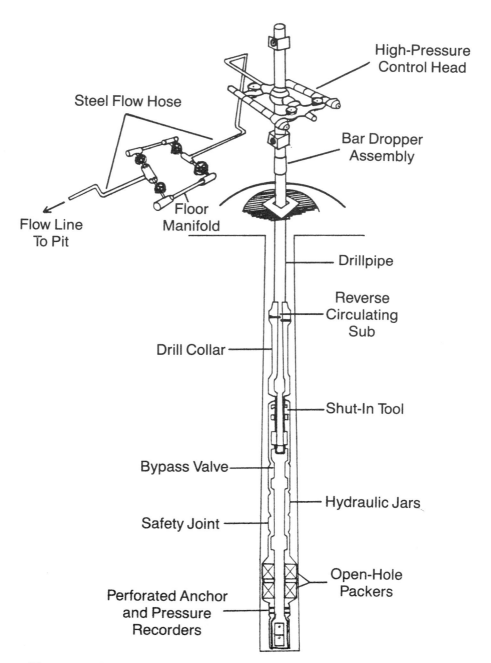

Fig. 10–5 Schematic of tools used in a drillstem test. Open-hole packer seals the formation from the mud column and opens a flow path to the surface for formation fluids through the drillpipe.

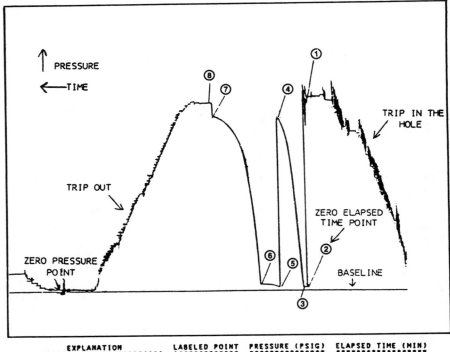

Fig. 10-6 Pressure buildup curve. See legend for interpretation.

EXPLANATION	LABELED POINT	PRESSURE (PSIG)	ELAPSED TIME (MIN)
HYDROSTATIC MUD	1	3592	-2.5
START FLOW	2	18	0.0
END FLOW & START SHUT-IN	3	23	3.6
END SHUT-IN	4	3215	37.2
START FLOW	5	19	36.7
END FLOW & START SHUT-IN	6	84	61.1
END SHUT-IN	7	3252	121.9
HYDROSTATIC MUD	8	3497	126.9

Another drawback is that the results may be ambiguous in the following situations:

1. If no fluid is recovered, the packer might not be set correctly or formation damage (clay swelling) might be the problem.
2. If only mud filtrate is recovered, very deep invasion may be the answer. The zone may be productive or it may be wet.
3. If drilling mud is recovered, the packer might not be making a good seal against the formation.
4. If a long interval is tested, the location of the oil or gas might be difficult to pinpoint.

Wireline Formation Tests

Formation tests may also be made with a wireline tool. The theory of the wireline tester is simple: An empty chamber or "jug" is lowered until it is opposite the zone to be tested. The chamber has an opening so fluids can flow into it. A valve controls the opening, and a pressure gauge measures the formation pressures. The opening in the chamber is in the center of a rubber pad that is pressed against the formation. This pad seals out the mud in the wellbore and allows only formation fluids to enter the chamber.

Once the seal is made, the tool is opened and fluids flow in. The flowing pressure is measured during this time. When the chamber is full, the pressure builds up until it approaches reservoir pressure. The tool is closed, the seal to the formation is broken, and the tool is pulled back to the surface. The contents of the sample chamber are measured and analyzed, and the pressure buildup curve is used to calculate permeability and reservoir pressure.

Most tools today can be reset as often as needed to form a seal or to measure pressure only. Samples are generally limited to one or two trials before the tool is pulled from the hole.

The wireless test tool is limited because the sample size is so small. Also, the permeability that is measured is the permeability very close to the wellbore, which may not be representative of conditions farther away. Nonetheless, the wireline tester can be of great value when several zones must be tested, when a reservoir pressure measurement is needed (to check for depleted reservoir, for example), or when permeabilities are fairly high and a full sample chamber is likely.

THROUGH-DRILLPIPE LOGGING

Sometimes engineers can't lower conventional logging tools to the bottom of the hole. The reasons might include swelling shales that bridge off (block or plug) the hole, ledges, poor mud properties, lost-circulation zones, or high-pressure gas zones. For whatever reason, no conventional open-hole logs are obtained. In this case there are only two alternatives: to run casing without seeing any logs (and then run a porosity device through the casing) or to run whatever logs are possible through the drillpipe. Usually, engineers decide to run the through-drillpipe logs.

Through-drillpipe logging is a technique rather than a tool. The drillpipe is first run in the hole open-ended (without a bit). The pipe

Table 10-1 Through-Drillpipe Logging Tools

Tool	Diameter, in.	Measurement
Induction electric	2 3/4	Resistivity
Density	2 3/4	Porosity
Compensated neutron	2 3/4	Porosity
Sonic	1 11/16	Porosity, waveforms
Temperature	1 11/16	Temperature
Gamma ray	1 11/16	Correlation, lithology
Thermal decay tool	1 11/16	Porosity, sigma

is lowered until it has passed the area that was causing trouble. Small-diameter (2 3/4 in. maximum) logging tools are run through the drillpipe and into the open hole below the drillpipe (Table 10-1). The logs are then recorded in the open-hole section below the drillpipe.

These logging devices are limited in the variety of measurements available and often in the sophistication of the measurement. The resistivity measurement is usually a deep induction curve and an old-style short normal curve. Although the density tool is compensated for mudcake, it does not have a lithology curve (photoelectric absorption index) or a caliper measurement.

Pulsed Neutron Logging

This logging device is used primarily in casing. Some tools make both a porosity and a sigma measurement. (All of them make the sigma measurement.) Sigma response is very similar to the conductivity of the formation, i.e., high sigma corresponds to high water saturations.

Since the tool can measure through steel, it is often useful in cases where logs were not obtained in the open hole. The log can be run either through the drillpipe or after the casing is in place.

However, there are limitations. First, the tool makes a very shallow measurement and is greatly affected by the invaded zone. If the tool is run through the drillpipe, only S_{xo} will be determined. If it is run immediately after a well has been cased and cemented, the tool will see only the mud filtrate and will still measure only S_{xo}. Several weeks must pass to allow the mud filtrate to disperse into the formation.

Second, the porosity is affected by the presence of gas in the formation. Gas makes all neutron porosities appear too low. Since no other porosity tools read reliably through casing, it is impossible to verify the porosity measured. In oil- and water-bearing formations the porosity will be correct if the lithology is known.

Specialty Logs 173

A third limitation of the tool is porosity. Reliable water saturations cannot be obtained if the porosity is less than about 10%.

A final limitation is that the formation water salinity must be in the range of 50,000 parts per million (ppm) chlorides; otherwise, there is not enough contrast between formation water and hydrocarbons to allow an accurate calculation of water saturation.

MEASUREMENT WHILE DRILLING

Some people have called MWD (measurement while drilling) technology "the success story of the '80s."* Today these tools are used extensively offshore; onshore they are becoming increasingly competitive with wireline logging techniques in some instances.

The basic MWD system is a collection of tools, built up as groups of modules on the primary steering tool (Fig. 10–7). It usually provides gamma-ray, resistivity, and directional data acquired while drilling.

The data is transmitted back to the surface by mud pulses which move at 4,100 ft/sec and are decoded by a computer on the surface. The changes in pulsation are interpreted by the computer as readings about the formations being drilled through. In some instances, the information is not relayed through the mud but is retained in a downhole computer and recovered when the drillstring is pulled from the hole.

CASED-HOLE LOGS

All of the logs we've talked about so far have had at least one thing in common: they are open-hole logs. They are run before the well is cased and are part of the original evaluation of the well. Another type of log is run in the well after the well is closed. These logs, generally called cased-hole logs, are subdivided into completion logs, production logs, and through-casing evaluation logs.

Completion Logs

If a well is believed to be productive, or at least to have a good chance of being productive, the operator will decide to run pipe. After the casing is run into the wellbore, it is cemented into place. The cement

---

*Paul Dempsey, *World Oil*, July 1987.

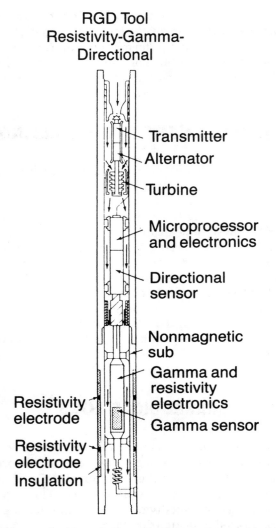

Fig. 10–7 MWD tools. Note that many modules are added to the primary steering tool.

serves three functions. First, it must support the weight of the casing string; second, it must effectively isolate the productive interval from the nearby formations; third, it must be strong enough to contain the pressures from hydraulic fracturing or acidizing operations (well stimulation). To evaluate whether the cement meets these criteria, one of several logs may be run.

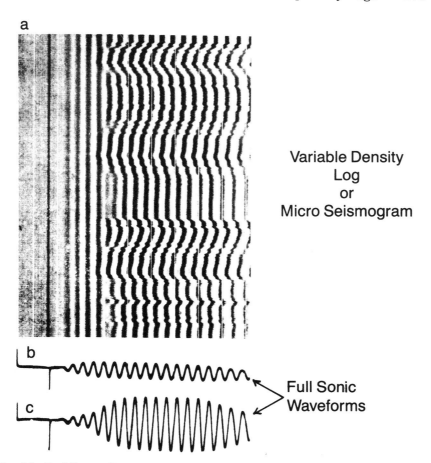

Fig. 10–8 Microseismogram (a) and CBL waveforms (courtesy Welex). (b) Well-cemented casing; (c) uncemented pipe.

The most common log used to evaluate the cement job is the cement bond log (CBL). This tool is an adaptation of the sonic tool that we looked at earlier. It measures not only the arrival time of the sound wave but also the amplitude of the first arrival of a sound wave that has traveled through the casing. If a solid material, such as cement, is in contact with the outside of the pipe, the sound is deadened and the amplitude is attenuated, or reduced (Fig. 10–8). If only liquid lies behind the casing, the pipe rings like a bell and the sound wave is not attenuated. To help interpret the amplitude curve, the full sonic wave train is generally recorded either as a series of waveforms or as a black-

and-white series of lines called a variable density log. In addition, the total transit time curve from the transmitter to the first receiver is recorded.

One of the limitations of the cement bond log is its inability to identify channels, mud-filled paths within the cement that can allow fluids to flow between nearby formations. The CBL is also affected by dense, low-porosity formations (fast formations) and by swelling of the casing due to pressure changes (microannulus effect).

A new generation of tool is being run more often to evaluate the cement. Schlumberger's tool is the Cement Evaluation Tool (CET); Gearhart calls its device the Pulse Echo Tool (PET). These tools have eight individual transducers that determine the radial distribution of the cement; any channels in the cement are readily apparent (Fig. 10–9). This ability to identify channels is the real strength of the new cement logs.

The temperature log is occasionally used to locate the cement top. Cement gives off heat as it sets up; the temperature log shows where this heating effect begins and so identifies the top of the cemented interval. It does not evaluate the quality of the cement job. The log must be run within 48 to 72 hr after the cement job is completed or the temperature difference will disappear.

A gamma-ray log is almost always run whenever a cased-hole tool is run in the hole. The gamma-ray curve is only slightly affected by the presence of the casing and therefore is an excellent way of tying the cased-hole depth measurements to the open-hole formation depths. This procedure is called correlating to the open-hole logs. The neutron tool also reads through casing and responds primarily to porosity, making it a good cased-hole correlation tool.

Once a cement has been evaluated and is found to be good, the next step (as far as wireline work is concerned) is to shoot the well. A perforating gun is lowered into the well; specially designed explosive charges are used to make holes in the casing, cement, and formation so the fluids can flow into the casing.

Production Logs

Once the well has been perforated and perhaps stimulated by acidizing or hydraulic fracturing, it is put on production. Occasionally some problems with production develop, such as unexpected water production, flow rates that drop off more rapidly than expected, or sudden changes in production. To evaluate the problem and possibly define a solution, production logs are often run.

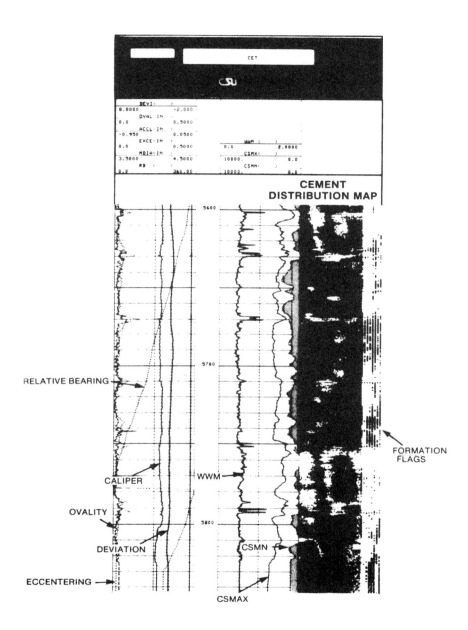

Fig. 10–9 Cement evaluation tool. Note radial distribution of cement around the casing.

These logs are usually run as combinations of measurements because the problems are generally complex. Some of the measurements possible are downhole flow rate; fluid density; temperature; pressure; noise; tracing the path of injected radioactive fluids; casing thickness, diameter, and corrosion; and natural radioactivity (gamma ray). From some combination of these measurements, engineers can often deduce what is happening downhole. With this information, remedial work may be instigated, such as setting plugs, moving packers, or opening new zones.

Through-Casing Evaluation

A third type of cased-hole log is through-casing formation evaluation. Although casing interferes with traditional evaluation measurements such as resistivity, there are devices that will read through the casing. We have discussed some of these already, including the compensated neutron log and the gamma-ray log. Under some conditions (such as excellently bonded casing), it is possible that sonic porosities and sonic waveforms can be recorded.

However, the pulsed neutron tools are the mainstay of through-casing evaluation. This category of tool measures either sigma (the macroscopic capture cross-section of the formation) or the ratio of carbon atoms to oxygen atoms. Both of these measurements give us water saturation if we know lithology and porosity. The logs are used either to monitor production by noting changes in water saturation over time or to look for hydrocarbon zones in old wells that may have been missed or disregarded when the well was originally drilled.

The logging industry is dynamic. Millions of dollars are spent annually in tool research and development. The results of this research make it to the "field" regularly in the form of new or better measurements.

AFTERWORD

The logging industry has come a long way from that day in 1927 when two French brothers took some electrical measurements on an oil well in Pechelbronn, France. The early "logging trucks" had cable made of telephone wire wrapped in friction tape. The winches were turned by hand, and at every revolution of the drum a bell rang. The bell signaled the helpers to stop turning the winch; then the engineer made a stationary measurement of voltage, which he converted to resistivity and recorded on a graph. These were the first logs.

From these very simple measurements, logging sophistication has advanced to space-age electronics, where spectroscopic measurements and downhole microprocessors are the norm. Logging trucks now carry enough steel-armored cable to log wells nearly 6 miles deep with bottom-hole temperatures over 450° F and pressures greater than 20,000 psi. Nearly 100 different measurements are available, from resistivity to depth to sonic waveforms to permeability.

The advent of the computer has caused a data explosion that is still growing. More ways of handling, presenting, and understanding the wealth of information available to us through logs are being developed daily. Where it will end is anybody's guess—perhaps with your computer calling a service company's computer and saying it has a well to log.

NOMENCLATURE

SYMBOLS

Traditional Symbols	Standard SPE of AIME and SPWLA[a]	Standard Computer Symbols[a]	Description	Customary Units or Relation
a	K_R	COER	coefficient in $F_R - \phi$ relation	$F_R = K_R/\phi^m$
C	C	ECN	conductivity, electric	millimhos per meter (mmhos/m)
C_p	B_{ep}	CORCP	sonic compaction correction factor	$\phi_{SVcor} = B_{ep}\phi_{SV}$
D	D	DPH	depth	feet (ft) or meters (m)
d	d	DIA	diameter	inches (in.)
F	F_R	FACHR	formation resistivity factor	$F_R = K_R/\phi^m$
H	I_H	HYX	hydrogen index	
h	h	THK	thickness (bed, mud-cake, etc.)	feet, meters, inches
K	K_e	COEC	electrochemical SP coefficient	$E_c = K_c \log(a_w/a_{mf})$

Symbol	Code	Description	Units
...	...	porosity (cementation) exponent	$F_R = K_R/\phi^m$
n	SXP	saturation exponent	$S_w^n = F_R R_w/R_t$
p	PRS	pressure	pounds/sq. inch (psi), kilograms per sq cm^c, atmospheres
R	RES	resistivity	ohm-meters (ohm-m)
r	RAD	radial distance from hole axis	inches
S	SAT	saturation	fraction or percent of pore volume
T	TEM	temperature	degrees (°F or °C), or kelvin (K)
BHT, T_{bh}	TEMBH	bottom-hole temperature	(same units as Temperature)
FT, T_{fm}	TEMF	formation temperature	(same units as Temperature)
t	TIM	time	microseconds (microsec), seconds (sec) minutes
v	VAC	velocity (acoustic)	feet per second, meters per second
V	VOL	volume	cubic centimeters (cc), cubic feet, etc.
V	VOL	volume fraction	
ϕ	POR	porosity	fraction or percentage of bulk volume, porosity unit (p.u.)
Δt (script t)	TAC	Sonic interval transit time	microseconds per foot
ρ (rho)	DEN	density	grams per cubic centimeter (gm/cc)
Σ (sigma)	XST	neutron capture cross section	capture units (c.u.), reciprocal

References: Supplement V to 1965 Standard—"Letter and Computer Symbols for Well Logging and Formation Evaluation," in *Journal of Petroleum Technology* (October 1975), pages 1244-1261, and in *The Log Analyst* (November-December 1975), pages 46-59.

SUBSCRIPTS

Traditional Subscripts	Standard SPE of AIME and SPWLA[a]	Standard Computer Subscripts[a]	Explanation	Example
a	LOG	L	apparent from log reading (or use tool-description subscript)	R_{LOG}, R_{LL}
a	a	A	apparent (general)	R_a
b	b	B	bulk	ρ_b
bh	bh	BH	bottom hole	T_{bh}
cor, c	cor	COR	corrected	t_{cor}
cp	cp	CP	compaction	B_{cp}
dol	dol	DL	dolomite	t_{dol}
e, eq	eq	EV	equivalent	R_{weq}, R_{mfeq}
f, fluid	f	F	fluid	ρ_f
fm	f	F	formation (rock)	T_f

Subscript		Description	Symbol
gxo	GXO	gas in flushed zone	S_{gxo}
h	H	hole	d_h
h	H	hydrocarbon	ρ_h
hr	HR	residual hydrocarbon	S_{hr}
i	I	invaded zone (inner boundary)	d_i
irr	IR	irreducible	S_{wi}
j	J	liquid junction	E_j
log	LOG	log values	t_{LOG}
ls	LS	limestone	t_{ls}
m	M	mud	R_m
max	MX	maximum	ϕ_{max}
ma	MA	matrix	t_{ma}
mc	MC	mud cake	R_{mc}
mf	MF	mud filtrate	R_{mf}
mfa	MFA	mud filtrate, apparent	R_{mfa}
min	MN	minimum value	
o	O	oil (except with resistivity)	S_o
or	OR	residual oil	S_{or}
o, 0 (zero)	ZR	100-percent water saturated	F_0

Continued.

SUBSCRIPTS—cont'd

Traditional Subscripts	Standard SPE of AIME and SPWLA[a]	Standard Computer Subscripts[a]	Explanation	Example
r	r	R	relative	k_{ro}, k_{rw}
r	r	R	residual	S_{or}, S_{hr}
sd	sd	SD	sand	
ss	ss	SS	sandstone	
sh	sh	SH	shale	V_{sh}
SP	SP	SP	spontaneous potential	E_{SP}
SSP	SSP	SSP	static SP	E_{SSP}
t, ni	t	T	true (as opposed to apparent)	R_t
T	t	T	total	C_t

xo	xO	flushed zone	R_{xo}
0 (zero)	ZR	100-percent water saturated	R_o
IL	I	from Induction Log	R_I
ILd	ID	from Deep Induction Log	R_{ID}
ILm	IM	from Medium Induction Log	R_{IM}
LL	LL	from Laterolog. (Also LL3, LL7, LL8, LLd, LLs.)	R_{LL}
(Also LL3, LL8, etc.)			
6FF40		from 6FF40 IL	R_{6FF40}
MLL	MLL	from Microlaterolog	R_{MLL}
PL	P	from Proximity Log	R_P
N	N	from normal resistivity log	R_N
D	D	from Density Log	ϕ_D
N	N	from Neutron Log	ϕ_N
SNP	SN	from Sidewall Neutron Log	ϕ_{SN}, ϕ_{SNP}
CNL	CN	from Compensated Neutron Log	ϕ_{CN}
TDT	PNC	from Thermal Decay Time Log	ϕ_{pnc}, ϕ_{TDT}
S	SV	from Sonic Log	ϕ_{SV}
ND		from Neutron and Density Logs	ϕ_{ND}
GR	GR	from Gamma Ray Log	ϕ_{GR}

References: Supplement V to 1965 Standard—"Letter and Computer Symbols for Well Logging and Formation Evaluation," in *Journal of Petroleum Technology* (October 1975), pages 1244-1261, and in *The Log Analyst* (November-December 1975), pages 46-59.

GLOSSARY

acoustic tool   *see* Sonic Log.
bulk volume   the amount of substance that is present within a unit volume; abbreviated BV. *See* Unit Volume. Bulk volume is expressed as a percentage of the unit volume. If all individual bulk volumes are added together, they must equal 100%.
bulk volume gas   the percentage of the unit volume that is gas; abbreviated BVG.
bulk volume hydrocarbon   the percentage of the unit volume that is hydrocarbon (includes both oil and gas); abbreviated BVH.
bulk volume matrix   the percentage of the unit volume that is non-porous rock; the mineral structure; abbreviated BVM.
bulk volume oil   the percentage of the unit volume that is oil; abbreviated BVD.
bulk volume water   the percentage of the unit volume that is formation water; abbreviated BVW.
bulk volume water—minimum   the BVW at which the formation water is bound to the formation grains by capillary forces; the same as BVW_{irr}. At BVW_{min} no water will be produced from the formation.
capture   a type of reaction that occurs when a neutron is absorbed into or captured by an atom. As this happens, the atom becomes highly energized and releases the energy by emitting a gamma ray.
clastic   rock formed from the fragments of other rocks.
compensated neutron log   tool that measures in the same way as a neutron log but also compensates for hole rugosity, measures ratio of detector responses, and converts ratio to a linear porosity reading.
compressional wave   type of sound wave that travels by compressing the material in which it travels; also called a primary wave or P-wave.
Compton scattering   condition in which a gamma ray hits an electron and imparts some of its energy to that electron.
conglomerate   a poorly sorted clastic sediment with large grains.
contamination gas   gas introduced into the drilling fluid from a source other than the formation.
correlation scale   vertical scales of 1 to 2 in./100 ft which geologists use to compare several wells over large intervals of formation.
cut   leaching of oil from a sample by a solvent.
density   weight of a unit volume of material divided by the weight of the same volume of water.
depth track   column down the center of a log that records the depth of the well in multiples of 100 ft.
detail scale   vertical scale of 5 in./100 ft which can be used to detect more features than on a standard 1- or 2-in./100 ft scale.
drainage area   the area that a reservoir covers or that a well can drain.

driller's log record of what occurs on a drilling rig, recorded by depth; notes types of rock encountered, rate of drilling, oil or gas flows, equipment breakdowns, accidents, and any other occurrence that might have a bearing on evaluating the well.

drilling rate curve rate of penetration expressed in units of length per hour.

drilling time curve rate of penetration expressed in minutes per unit of length.

dual induction laterolog tool developed for areas of low-to-moderate resistivity and deep invasion; *see* Induction Log. Has two induction curves (deep and medium) plus a shallow-reading laterolog curve.

dual laterolog tool developed for areas of high resistivity and deep invasion, has deep and medium laterolog curves. *See* Laterolog.

effective porosity pore space available to hold fluids.

electric log tool that emits current from a constant electrode source and then measures the current at another electrode some distance away, with respect to a reference electrode. The earliest form of resistivity log.

electric wireline wire rope with insulated electrical wires or conductors beneath the strands of cable.

evaporites rocks formed from precipitate residue after a salty body of water evaporates.

flushed zone area where filtrate has flushed out all original fluids possible; abbreviated xo.

focused electric log tool used on highly resistive formations; the measuring current is forced into the formation by guard electrodes.

fracture porosity porosity attributed to fracture planes rather than to pores; the porosity is low, but the permeability is high.

gamma-ray log tool that reads gamma rays naturally emitted from formations.

gas chromatography method of analyzing the composition of the gas stream on a regular but intermittent basis.

ground loops circular currents generated by an induction tool that are concentric with the tool axis.

header the top of the log, where well data are noted.

hydrostatic pressure the pressure exerted by a column of fluids usually associated with the wellbore.

induction electric log a logging tool with a single induction curve combined with a short normal or shallow laterolog curve, used in medium- to high-porosity formations.

induction tool device that uses an alternating magnetic field to create ground loop currents in the formation concentric with the wellbore. The resistivity of the formation is inversely proportional to the amount of current induced in the formation.

intergranular porosity *see* Matrix Porosity.

invaded zone area including the flushed zone and extending into the formation to the depth that wellbore fluids have mixed with the formation fluids; denoted as subscript i.

invasion   a condition in which mud filtrate penetrates the formation next to the wellbore. The mud filtrate may cause swelling of the shale in the formation, which restricts production.

irreducible water saturation   the minimum possible water saturation in a formation; a fraction of grain size, surface area of the sand grains, and shaliness. A zone that is at irreducible water saturation will produce all hydrocarbons and no water.

kick   a condition in which the formation fluids flow into a well without being controlled. A major kick is called a blowout.

lag   the amount of time that elapses from the moment when the drill bit penetrates a new formation until the moment when the downhole particles and/or traces of gas circulate to the surface.

lateral   a specific arrangement of electrodes on an electric log. The lateral curve reads more deeply into the formation than a normal curve.

laterolog   a simple focused log with electrodes that force any current into the formation to minimize the effect of the low-resistivity borehole.

liberated gas   fluid released from exposed pores in a formation that mixes with the drilling mud and flows back to the surface.

lithology   rock type, such as sandstone, shale, limestone, dolomite, anhydrite, etc.

log   data recorded versus depth or time in graph form or with accompanying written notes. Also, a common term for any tool that is run downhole and generates or measures signals recorded on the printed log.

long normal   *see* Normal.

lost circulation   a condition in which large amount of drilling mud are pumped into the formation and lost in fractures or vugs.

matrix   the mineral structure from which a formation is made.

matrix porosity   pore spaces between rock grains whose nonrock volume equals the porosity.

microlaterolog   a smaller version of a laterolog, used in high-resistivity formations; used to measure R_{xo}, flushed zone resistivity; abbreviated MLL.

microlog   an early resistivity tool with a printout consisting of a caliper curve and two resistivity curves; most effective in low- to medium-resistivity formations. Used to indicate permeability.

microresistivity logs   tools designed to read the resistivity of the flushed zone; a very shallow-reading tool.

microspherically focused log   a pad device whose readings are recorded on a logarithmic scale. Useful in helping determine depth of invasion, S_{xo}, moved hydrocarbons, permeability, porosity, hole diameter, and zone thickness; abbreviated SMFL.

mudcake a sealant present in the drilling mud, used to protect the formation from invasion.

neutron log tool that bombards formations with neutrons from a radioactive source housed in the tool and measures the porosity as a function of the number of hydrogen atoms present.

normal a specific electrode arrangement on an electric log; often categorized as short normal or long normal, depending on the spacing.

offset well a nearby well whose logging data and production history are assessed during plans to drill a new well in an existing field.

ohms in logging terminology, an abbreviation for ohmmeters, which is an abbreviation for meters squared per meter. In the field, it is referred to as "ohms," although purists prefer the longer version.

Ohm's Law a principle which states that voltage is equal to current multiplied by resistance.

oil saturation the percentage of the pore space filled with oil; abbreviated S_o. If no gas is present, $S_w + S_o = 1$.

photoelectric effect condition that occurs when a low-energy gamma ray, passing close to an atom, is absorbed; an electron is then ejected into space.

porosity cutoff the minimum porosity established for production. Formations with lower porosity are not counted as net pay.

pressure differential the difference between the wellbore or hydrostatic pressure and the formation pressure.

primary wave *see* Compressional Wave.

rate of penetration the speed at which a bit penetrates a formation; abbreviated ROP.

recovery factor the percentage of oil or gas that may be recoverable by primary production; used in the reserve calculation.

refraction a characteristic of sound waves; their change in speed when the material through which they travel changes.

reserves the amount of recoverable oil or gas in a formation.

residual oil oil that remains in the formation and which cannot be removed by normal means. Residual oil saturation is abbreviated S_{or}.

resistance equal to resistivity times length through which current flows divided by the cross-sectional area; $r = \rho L/A$.

resistivity a physical parameter or property of a material. A measure of the difficulty an electric current will encounter in passing through the material. The units are ohms meters^2/meter or ohmmeters.

resistivity profile the difference in readings between the shallow, medium and deep resistivity curves.

rugosity roughness

secondary wave *see* Shear Wave.

shale base line a fairly constant reading opposite shales, used as a guide on the spontaneous potential log.

Glossary 191

shale bound water volume water that is chemically bound to the shale and is not free to move.

shear wave a sound wave slower than a compressional wave which moves vertically rather than horizontally to the axis of the wave; also called an S-wave or secondary wave.

short normal *see* Normal.

sidewall neutron porosity tool a device with detectors mounted on a skid that is pressed against the borehole wall, generally used in air-drilled holes; abbreviated SN.

sonde synonym for logging tool.

sonic log a tool that uses sound waves to measure porosity in a formation.

spontaneous potential a naturally occurring voltage caused when conductive drilling mud contacts the formations; abbreviated SP.

stacking the practice of connecting logging tools so they can be run simultaneously.

static SP the maximum spontaneous potential that can be measured if no current flows in the borehole; a constant reading. Abbreviated SSP.

stroke counter a mechanical device used to count the strokes of the reciprocating mud pumps. Since the volume of mud that the pump moves with each stroke is known, the rate and volume of mud pumped can be determined by counting the strokes. The hole volume and the displacement of the drilling string can be accurately estimated. The number of strokes on the stroke counter necessary to circulate the cuttings to the surface can be calculated from this information.

thermal catalytic combustion also known as hot-wire detection (HWD), one of the main ways to detect gas in the drilling mud; abbreviated TCC.

total porosity percentage of nonrock volume in a rock; may be filled with oil, gas, formation water (including bound water), or shale. Porosity is often referred to as void space in the rock—a misnomer.

track vertical column, often numbered 1, 2, or 3, in which specific reading from a logging tool is recorded.

transmitter current high-frequency current generated by an induction tool.

true resistivity resistivity of an uninvaded or uncontaminated zone; abbreviated R_t.

unit volume a cube of formation one unit long (a unit can be an inch, a meter, or a mile—it's not important) on each side. The volume (V) of a unit volume is 1 unit × 1 unit × 1 unit = 1 unit.

vertical resolution the thinnest bed a logging tool will detect.

virgin zone undisturbed or uncontaminated formation.

vug a space in the rock that is larger than a pore, often ranging up to cavern size.

vugular porosity porosity attributed to caverns or vugs in the rock, usually caused by dissolution of the rock by water movement.

water saturation the percentage of the pore space that is filled with formation water. If a formation's porosity is completely filled with water, $S_w = 100\%$. If oil and water are present, $S_w + S_o = 100\%$.

weighting material any solid (such as barite) added to the drilling mud to raise the fluid pressure of the mud column.

SUGGESTED READING

For the reader who is interested in a more detailed discussion of logging tools, their application and interpretation, the following list of books and service company manuals should be helpful.

Essentials of Modern Open-Hole Log Interpretation, John T. Dewan, 1983, Pennwell Publishing Co.

Manuals commonly available from the service companies:

    Log Interpretation Manuel
        Includes theory, procedures, and applications

    Log Interpretation Charts

    Open Services Catalog

    Cased Hole Services Catalog

These manuals may be ordered by writing to the service company.

Altas Wireline Services
P.O. Box 1407
Houston, Texas 77251

Schlumberger Well Services
P.O. Box 2175
Houston, Texas 77023

Gearhart Industries
P.O. Box 40058
Fort Worth, Texas 76140

Welex Headquarters
P.O. Box 42800
Houston, Texas 77242-8032

INDEX

A

acoustic log; *see* Log
air-filled holes   101, 164-165
Archie equation   42, 117, 142
atoms   85-87

B

backup curves; *see* Curves
baseline   47, 77
bed thickness correction   137
bulk volume; *see* Volume

C

caliper curve; *see* Curves
capture   89
carbonates   26, 30, 58, 96
cased-hole logs; *see* Logs
casing   100
cementation exponent   127
cement log; *see* Logs
clastics   25
commercial production   3
compensated density log; *see* Logs
compensate neutron log; *see* Logs
completion engineer   4, 5
completion log; *see* Logs
compressional wave; *see* Waves
Compton scattering   87-89
computer center logs; *see* Logs
computer-generated logs; *see* Logs
conductivity   40
conglomerates   25
cores   83-85
correction charts   72-73
correlation scales; *see* Scales
cross-plot technique   106-113
crush cut   59
cumulative porosity-feet   116
curves; *see also* Logs
   backup   12-13
   caliper   80
   density   91-92
   density correction   91-92
   drilling rate   46
   gamma-ray   12-13, 113-114
   microinverse   80
   micronormal   80
   photoelectric index   91-92
   pressure buildup   168, 170
   resistivity   12-13, 118

D

definition of log   1
deflection   47-48
density   89-90
   matrix   90-91
   fluid   90-91
density correction curve; *see* Curves
density logs; *see* Logs
depth column; *see* Depth Track
depth track   9
detail scale; *see* Scale
development geologist   3, 4
dielectric constant log; *see* Logs
dip   161-162
dipmeter; *see* Logs
disclaimers; *see* Escape Clause
divisions   12-15
driller's log; *see* Log
drilling break   48, 60
drilling engineer   4, 5
drilling mud   35
drilling rate curve; *see* Curve
drilling responses   48-49
drillstem test   168-170
dual induction laterolog; *see* Logs

E

effective porosity; *see* Porosity
electric log; *see* Logs
electron   85-87
epithermal neutron log; *see* Logs
escape clause   xii
evaporites   25, 58, 94
exploration geologist   3, 4

Well Logging for the Nontechnical Person

F

flame ionization detector   51
flash cut   59
fluid volume; *see* Volume
fluorescence   57, 59
flushed zone   36-38
flushed-zone resistivity   64
focused electric log; *see* Logs
formation resistivity factor   40-41
formation
  thickness   81
  undisturbed   35-36
FR   14
fracture porosity; *see* Porosity

G

gamma ray   85-88, 89
gamma-ray curve; *see* Curves
gamma-ray log; *see* Logs
gamma-ray track; *see* Tracks
gas chromatography   50
gas detection   48-51
ground loops   67
guard log; *see* Logs

H

header   7-9
history   1, 63
horizontal scale; *see* Scales
hydrocarbon ratio analysis   60
hydrocarbon saturation   38
hydrogen atoms   100
hydrostatic pressure   35-36, 49

IJK

induction tools   65-70
infrared analyzer
inserts   16-19
intergranular porosity; *see* Porosity
invasion   33-38, 63-65
investors   4, 5
irreducible water   26-27
irreducible water saturation   35-38
isopach map   119-120

L

lag   48
landman   4, 5

lateral device   74
limestone   58, 78, 89, 91, 100
linear scales; *see* Scales
Litho-Density log; *see* Logs
lithology   45, 78, 87, 91, 94-98
lost circulation   35
logarithms   16-18
logging companies   9
logs; *see also* Curves
  acoustic   101-105
  cased-hole   173-178
  cement   176
  compensated density   85, 89
  compensated neutron   85
  completion   173-177
  computer center   153-158
  computer-generated   147-159
  density   89-99
  dielectric constant   165-166
  dipmeter   161-164
  driller's   1
  dual induction laterolog   68, 70
  electric   63, 66, 74-75
  epithermal neutron   165
  focused electric   70-73
  gamma-ray   100
  guard   72
  induction   65-70
  Litho-Density   91-93
  micro-   66, 79-81
  microlaterologs   66, 81
  microspherically focused   66, 72, 83
  mud   45-61, 121
  multiple porosity   105-111
  natural gamma-ray spectroscopy   166-167
  neuron-density   118
  noise   165
  production   176, 178
  pulsed neutron   172-173
  quick-look   148-151
  sidewall neutron porosity   101
  sonic   101-105
  spontaneous potential   76-78
  temperature   165
  wellsite   148-153

M

main log section   9-16
matrix   90
matrix density; *see* Density

matrix porosity; *see* Porosity
matrix volume; *see* Volume
magnetic fields   65
measurements   38-39
measurement while drilling   173
membrane potential   76
microinverse curve; *see* Curves
microlaterolog; *see* Logs
microlog; *see* Logs
micronormal curve; *see* Curves
microresistivity tools   78-81
microseismogram   175
microspherically focused log; *see* Logs
minimum bulk volume water
mud cake   35, 80
mud filtrate
mud log; *see* Logs
multiple porosity logs; *see* Logs

N

natural gamma-ray spectroscopy log; *see* Logs
neutron   85-87
neutron porosity; *see* Porosity
neutron tool   100
noise log; *see* Log
normal device   74

OPQ

pair production   87-89
permeability   29-30, 59, 80
photoelectric effect   86-89, 91
porosity
  core   83-84
  cross-plot   122
  effective   100
  fracture   30
  intergranular   26-29, 105
  matrix   30
  neutron   101
  sonic   105
  total   100-10
  vugular   30, 105
potential   76
pressure-buildup curve; *see* Curves
pressure differential   49-50
production log; *see* Logs
proton   85-87
pulsed neutron log; *see* Logs
P-wave; *see* Waves

quick-look logs; *see* Logs

R

rate of penetration
  interpreting   47-48
  measuring   45-46
reading logs   116-125, 127-146
readings, gas   52-54
receiver coil   67
recovery factor   128
reflection   102
refraction   102
reserves estimate   33, 139, 146
reserves number   128
reserves engineer   4, 5
residual oil saturation   37
resistivity   38, 39-43, 63-82, 116, 129
resistivity curves; *see* Curves
response   94-98
rugosity   100
running logs   2-3

S

samples   56-57
sandstones   25, 27, 30, 58, 77, 91
scales
  correlation   11, 116
  detail   116
  horizontal   11-16
  linear   9-10
  logarithmic   16-18
  super detail   11
  vertical   9-11
Schlumberger Well Log Co.   2
Schlumberger, Conrad and Marcel   63
sedimentary rocks   25-26
shale   25, 32, 58, 76-77, 98, 100, 113
shear wave; *see* Waves
shows   54, 57-61
sidewall neutron porosity log; *see* Logs
Simandoux equation   143
sodium ions   76
sonic log; *see* Logs
sonic porosity; *see* Porosity
spontaneous potential   74, 76-79
SP track; *see* Tracks
static SP   77-78
streaming cut   59
super detail scale; *see* Scales
S-wave; *see* Waves

T

tadpole plot   163-164
temperature log; *see* Logs
tests   57-59
thermal catalytic combustion   50
thermal conductivity detector   50
through-casing evaluation   178
through-drillpipe logging   171-173
tornado chart   68-69
total porosity; *see* Porosity
tracks
   gamma-ray   11
   spontaneous potential   11, 77
transition zone   38
transmitter current   65, 67

U

unit volume; *see* Volume

V

vertical resolution   68
vertical scales; *see* Scales
volume
   bulk   30-32, 34
   fluid   30-32, 34, 91
   matrix   30-32, 34, 91
   unit   30-31, 34
vugular porosity; *see* Porosity

W

water saturation   35-38, 143
waves
   compressional   101-102
   P-   101-102
   S-   102-103
   shear   102-103
well information   7
wellsite computer logs; *see* Logs
wireline   63
wireline formation tests   171

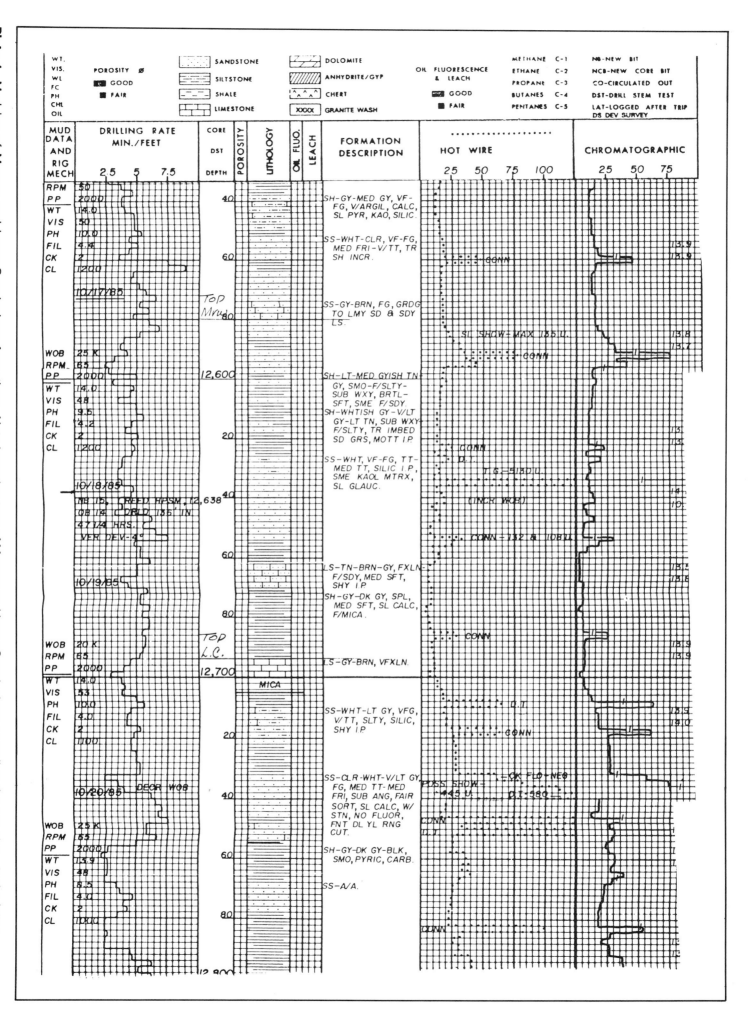

Plate 1 Mud log of Morrow and lower Cunningham formations. Note formation tops of Morrow and lower Cunningham as well as the gas show in the latter.

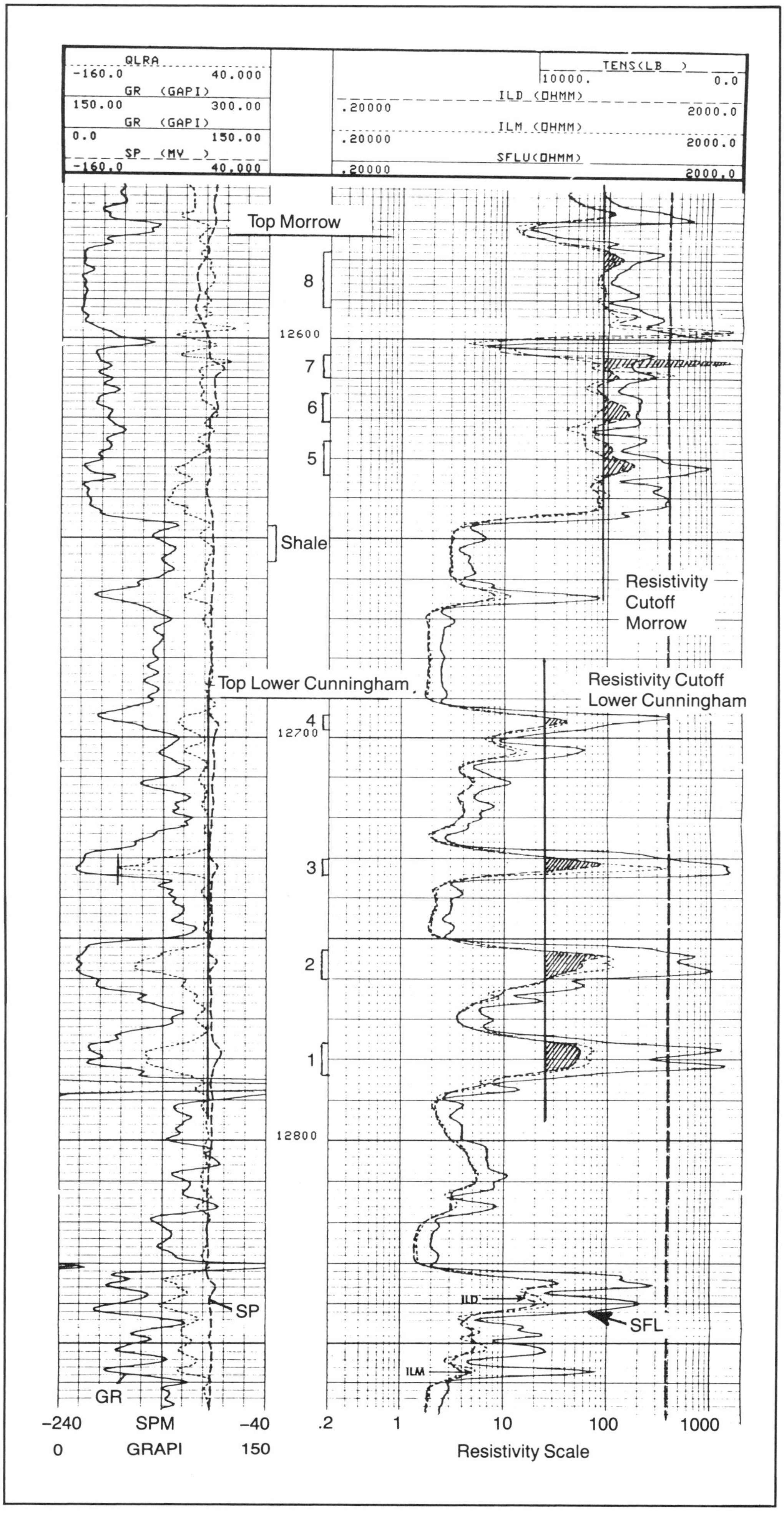

Plate 2 Dual induction log for example well. Note resistivity cutoffs in Morrow sand and lower Cunningham.

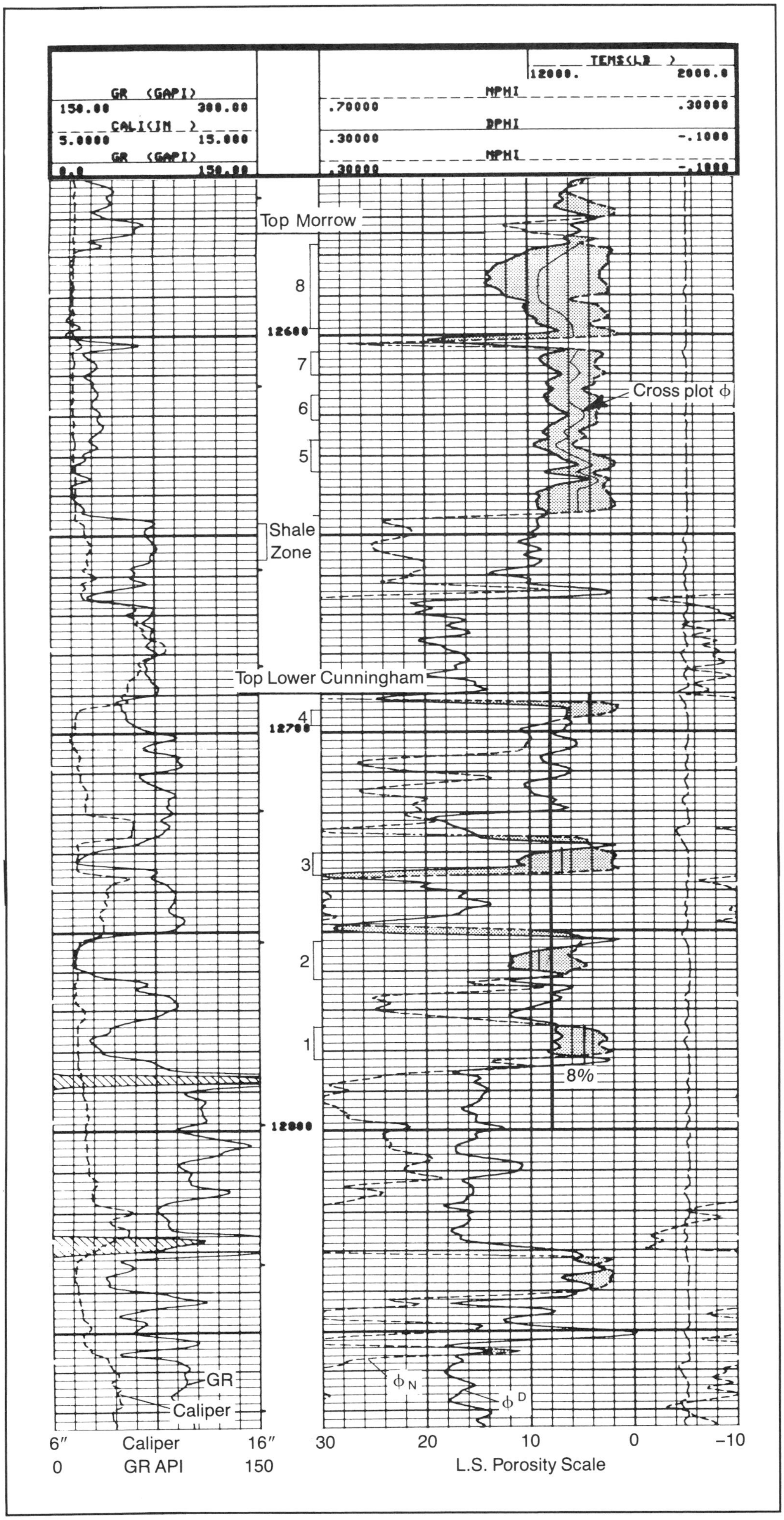

Plate 3 Compensated neutron-density log for example well. Note porosity cutoffs and cross-plot porosity drawn on the log.

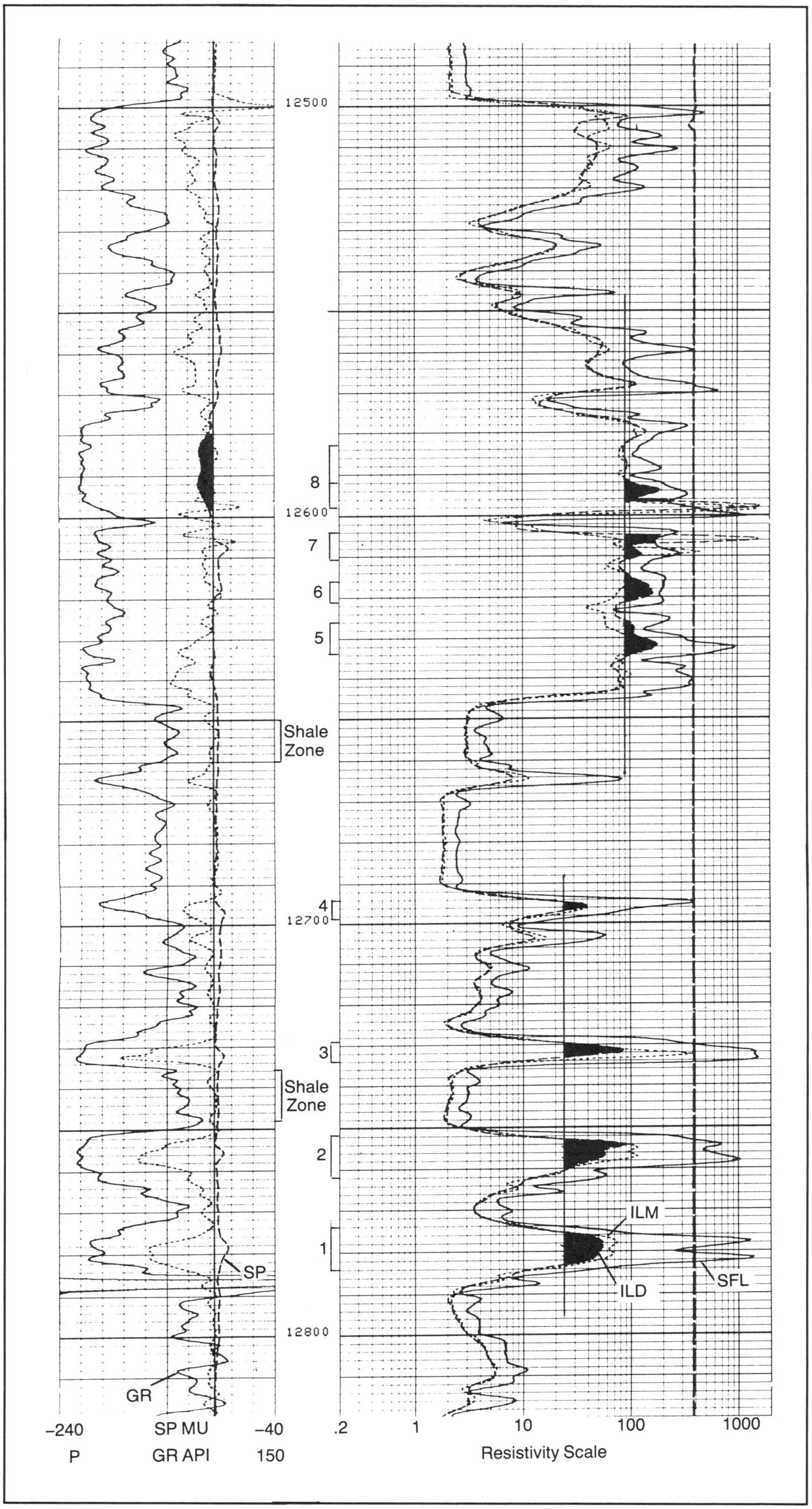

Plate 4 Dual induction log, Sargeant 1-5 well.

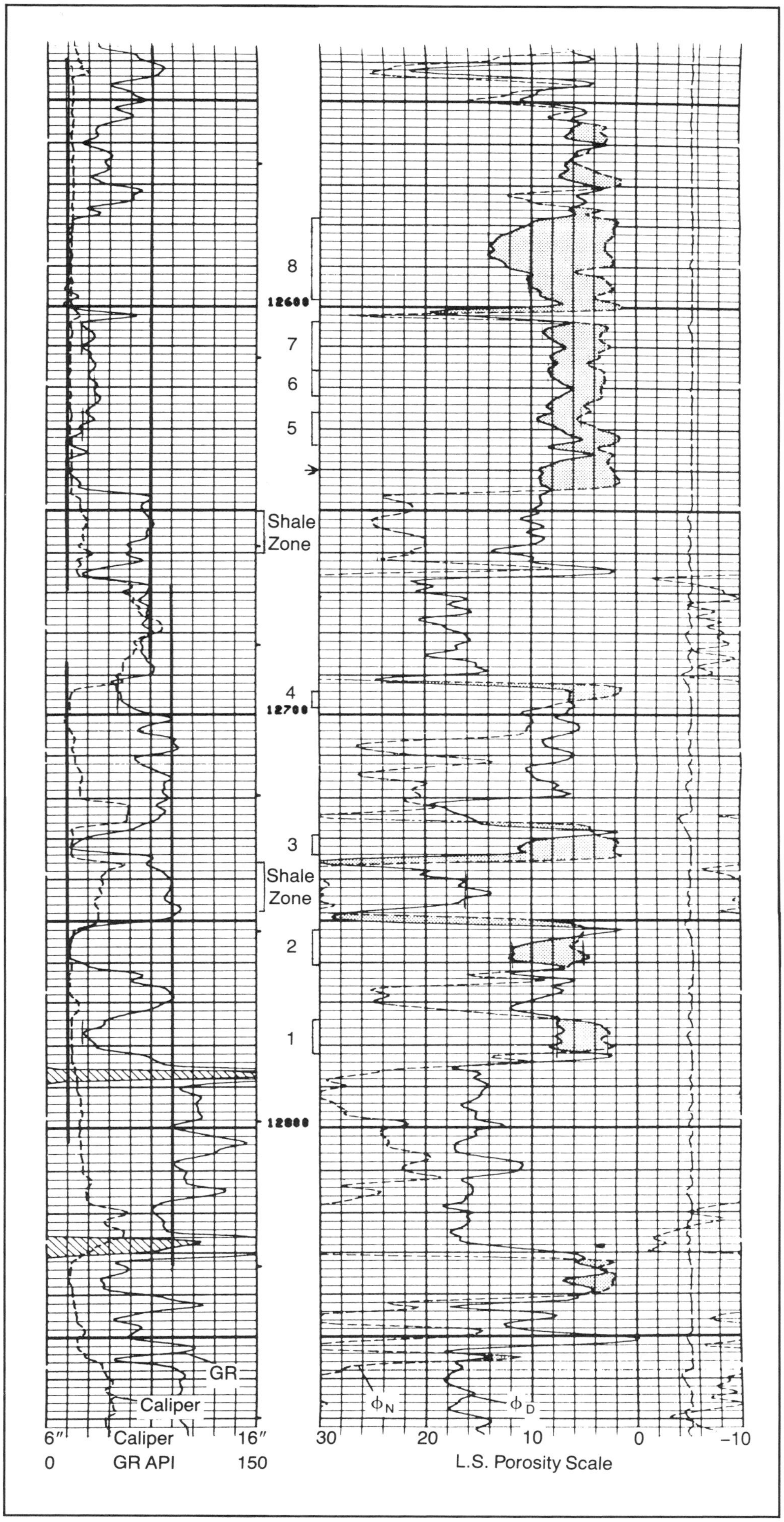

Plate 5 Neutron-density porosity log, Sargeant 1-5 well.

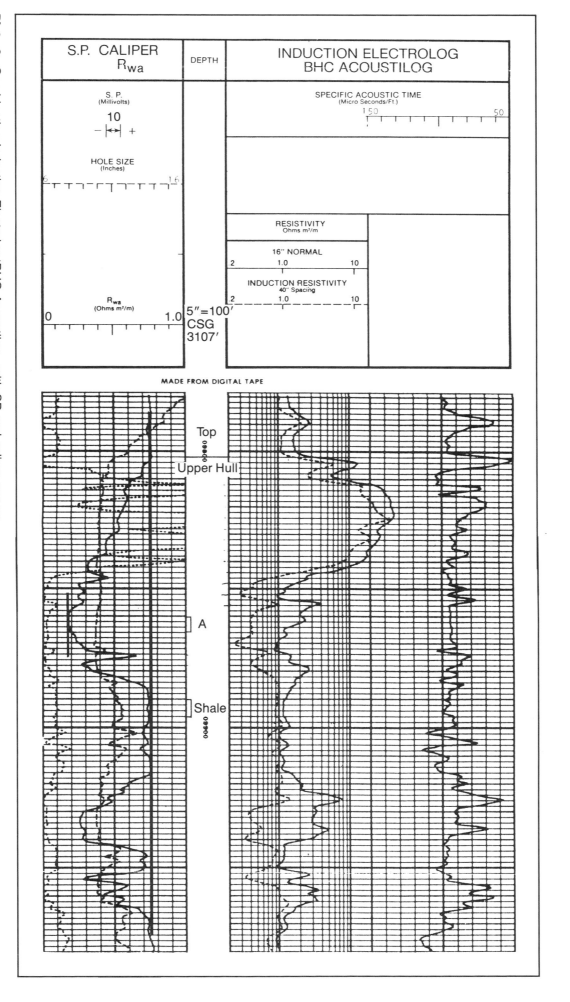

Plate 6 Combination Induction Electrolog/BHC Acoustilog_{tm} with SP and caliper curves.

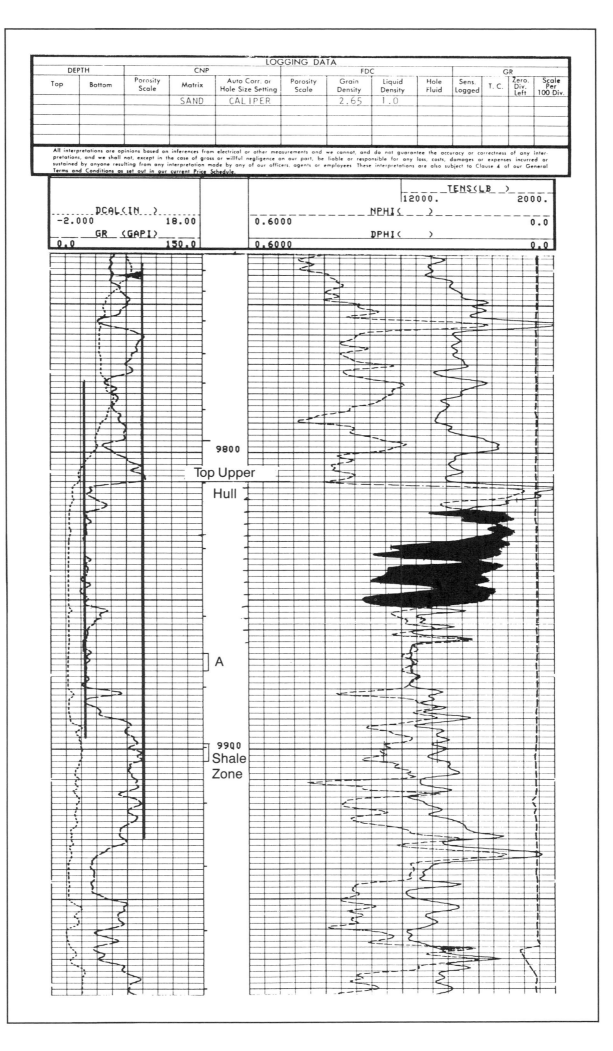

Plate 7 Compensated neutron-density porosity log with gamma-ray and caliper curves.